D0212953

Control and Dynamics in Power Systems and Microgrids

Control and Dynamics in Power Systems and Microgrids

Lingling Fan

CRC Press is an imprint of the
Taylor & Francis Group, an **informa** business

CRC Press
Taylor & Francis Group
6000 Broken Sound Parkway NW, Suite 300
Boca Raton, FL 33487-2742

CRC Press is an imprint of Taylor & Francis Group, an Informa business

Printed on acid-free paper
Version Date: 20170405

International Standard Book Number-13: 978-1-1380-3499-0 (Hardback)

Visit the Taylor & Francis Web site at
http://www.taylorandfrancis.com

and the CRC Press Web site at
http://www.crcpress.com

Contents

Preface

The main part of the book was written in Spring 2016 when the author taught Power Systems II, a graduate course at University of South Florida, Tampa Florida. The course was designed to focus on control and dynamics of power systems. Bergen and Vittal's book, *Power System Analysis*, was adopted as the textbook. Dynamics and control, especially in the area of power systems applications where rotating magnetic fields are involved, is a formidable subject to students. Hence, a set of class notes was developed in that semester to offer a tutorial approach of learning. Many examples and codes were developed to facilitate understanding and hands-on training. A highlight of this textbook is its many tutorial examples.

The first version of Bergen and Vittal's book was written in 1981 by Professor Bergen. Thirty-six years have passed since then. This classic textbook has been highly recognized and helped to educate a generation of power systems engineers. Professor Bergen passed away in July 2014. As a power systems engineer, this author would like to contribute to the field by reinterpreting the classics of power system control and dynamics. This textbook is also a tribute to Professor Bergen.

There will be several things different from the classic textbook.

The generator model derivation is very sophisticated in Bergen and Vittal (2009). In Bergen and Vittal (2009), Park's transformation was employed to derive generator models. The alternative of Park's transformation is space vector and complex vector transformation, a concept used much more often in machines and power electronics after the 1980s. The space vector concept makes Park's transformation straightforward. In this textbook, the author will explain synchronous generator dynamics, the most formidable dynamics in power systems, using space vector concepts.

In the 1980s, power electronics and microgrids were yet to be developed. This field is well developed in the 21st century. Many techniques used in power systems for power sharing, e.g., droop control, can also be found in power electronic converter coordination. This part is now related and put into the textbook to help readers understand converter control and coordination. This is another highlight of the textbook.

Advanced controls such as networked control (consensus control) were developed after 2000. Many classic engineering implementations follow the advanced control framework. It is appealing to find them and interpret the intuitive engineering design with concepts and ideas from networked control. In this textbook, inter-area oscillations are explained using consensus control.

Professor Bergen's book has steady-state analysis and dynamics all together. This textbook focuses on dynamics and control only. This author would also like to have a better flow to focus on power system control. Starting from the beginning, the ordinary differential equation, the building block of dynamics and control, is explained using examples. Dynamic simulation and linear system analysis are conducted for the examples. With the fundamental concept of dynamics built, readers can then pursue the learning tasks related to power system control and dynamic stability with ease.

The flow of the text is to treat frequency or voltage control as control problems. For control problems, first we discuss the plant model and its related steady-state and dynamic responses. The plant model should be identified with the inputs and outputs specified. In the frequency control case, it is obvious that the output of the plant model should be frequency. The inputs are from a generator's mechanical system inputs. After setting up the plant model, we then think about how to design feedback controls to realize control objectives. After the control design is conducted, we then employ dynamic simulation to verify controller performance.

The author is grateful to have the opportunity to write and publish this book through the CRC press. The author would like to acknowledge the University of South Florida Electrical Engineering Department for providing a great environment for conducting research and teaching. The author wishes to acknowledge her family for their encouragement.

The book was developed from the author's class notes of Power Systems II for Spring 2016. Minyue Ma, a Ph.D. student, was the teaching assistant for that course and helped work out examples and homework problems for the class. A few students in the class, e.g., Abdullah Alassaf, highly complimented the class notes, which encouraged the author to contact the CRC press for publication. Yin Li, a Ph.D. student, built the MATLAB®/Simulink models used in Chapter 6 Frequency and voltage control in microgrids. Yangkun Xu, another Ph.D. student drew many figures for Chapter 3 and Chapter 5. Graduate students at the USF Power Systems Smart Grid Lab reviewed the book during the holiday season in December 2016. The author wishes to acknowledge Minyue Ma and Yin Li as reviewers. The author also wishes to acknowledge Yi Yang from Eaton Cooperation as a reviewer.

MATLAB® is a registered trademark of The MathWorks, Inc. For product information please contact:
The MathWorks, Inc.
3 Apple Hill Drive
Natick, MA, 01760-2098 USA
Tel: 508-647-7000
Fax: 508-647-7001
E-mail: info@mathworks.com
Web: www.mathworks.com

Chapter 1

Introduction

1.1 Why a new textbook?

A traditional power system dynamics and control book covers synchronous generator models (steady-state and dynamics), generator voltage control, power system frequency control, and power system transient stability. A typical textbook is Bergen and Vittal's book *Power Systems Analysis* (Chapters 6, 7, 8, 11, and 14). With today's smart grid industry, the following aspects need to be added in teaching and in textbooks.

1. How to carry out demonstration for power system dynamics and control.

To address this task, dynamic simulation of ordinary differential equation-based models and further programming implementation in software environment such as MATLAB® or Python should be covered. This part is usually not found in a traditional textbook. Rather, students have to go to another course or read another book on computing to learn how to conduct validation and demonstration. In this text, tutorial examples on programming and dynamic simulation will be provided and students can quickly manage to conduct validation through coding or MATLAB/Simulink.

2. How to carry out control design.

Classic control methods such as the root locus method are repeatedly used in Bergen and Vittal (2009). In the 1980s, MATLAB and its control toolbox were not yet popular. Therefore, Bergen and Vittal (2009) did not present examples related to MATLAB codes. This textbook will provide MATLAB examples for control design problems.

3. How to better explain rotating machines.

Generator model derivation is the most sophisticated part in Bergen and Vittal (2009). In Bergen and Vittal (2009), Park's transformation was employed to derive generator models. The alternative to Park's transformation is space vector and complex vector transformation, a concept used much more often in machines and power electronics after the 1980s. The space vector concept makes Park's transformation straightforward. In this textbook, the author will explain synchronous generator dynamics, the most formidable dynamics in power systems, using the space vector concept.

4. How are microgrids controlled?

In the traditional power system dynamics and control books, the focus is on synchronous generators. With the current industry where renewable energy, power electronics converters and microgrids arise, the related system-level dynamics and control should be covered. For example, when frequency control is discussed, it is very natural to extend the applications from large-scale power systems to microgrids where droop control is also used. Coverage on microgrid control is a highlight of this textbook.

In short, the aim of this textbook is to provide more insights using programming examples, state-of-the-art control design tools, and advanced control concepts to explain traditional power system dynamics and control. In addition, microgrid control will be covered as extended applications.

While reading this textbook, readers will get the chance of training in programming and control design. They will gain knowledge on dynamics and control in both synchronous generator-based power systems and power electronic converter enabled microgrids.

1.2 Structure of this book

The book is organized in eight chapters. The book has two main parts: control (frequency and voltage control) and dynamics (large-signal stability and small-signal stability). Before control problems are introduced, the validation tool: dynamic simulation, is examined in Chapter 2. Along with dynamic simulation, linear system analysis tools such as Bode plots, are also introduced.

There are four chapters related to control: Chapters 3-6. Frequency control and power sharing of synchronous generators are examined in Chapter 3. Electromechanical dynamics of a synchronous generator is considered while electromagnetic dynamics are not considered in Chapter 3. This treatment makes analysis concise with only critical dynamics included. After frequency control, voltage control is to be examined. To better explain voltage control, a detailed examination of a synchronous generator's model with electromagnetic dynamics is required. Therefore, Chapter 4 focuses on the derivation of synchronous generator models using the space vector concept. Both steady-state and dynamic models are presented in Chapter 4. Chapter 5 presents voltage control of synchronous generator.

Chapter 6 covers converter control and power sharing among converters in a microgrid. The materials presented in Chapter 6 have never been found in any textbook on power system control and dynamics. Chapter 6 first presents a single voltage source converter's control. Depending on its operation mode, a converter can either work in PQ control mode for grid-connected operation or in VF control mode for autonomous operation. With the fundamental control covered, droop control for power sharing among converters is then presented. This chapter gives many examples on control design and simulation-based validation.

Part II of the book focuses on dynamics. Two chapters are included. Chapter 7 focuses on large-signal stability. An example is transient stability of a synchronous generator. Chapter 8 focuses on small-signal stability. Three engineering problems are used as examples in this chapter: small-signal model derivation of a single-machine infinite-bus (SMIB) system for stability analysis, inter-area oscillation explanation using networked control theory, and torsional interactions in a synchronous generator. For each problem, linear system models are derived and linear system analyses are conducted.

This book provides many examples and tutorials to facilitate learning. Through the study of this book, readers can master the skill of linear system analysis and simulation-based validation. What's more, this book builds a bridge between traditional synchronous generator-based large-scale power system control and converter-based microgrid control.

Part I: Control

Chapter 2

Dynamic Simulation

2.1 Introduction

In this chapter, we will describe how to build validation testbeds for control and dynamic analysis using dynamic simulation or time-domain simulation. Take a simple RL circuit example shown in Figure 2.1. We would like to know the current $i(t)$. The voltage source $v_s = V_{DC}$ is assumed as known, e.g., 5 V. Initially, the circuit is open. Assume that at t_0, the switch turns on. We first establish an ordinary differential equation (ODE) to describe the circuit model.

$$V_{DC} = Ri(t) + L\frac{di(t)}{dt} \tag{2.1}$$

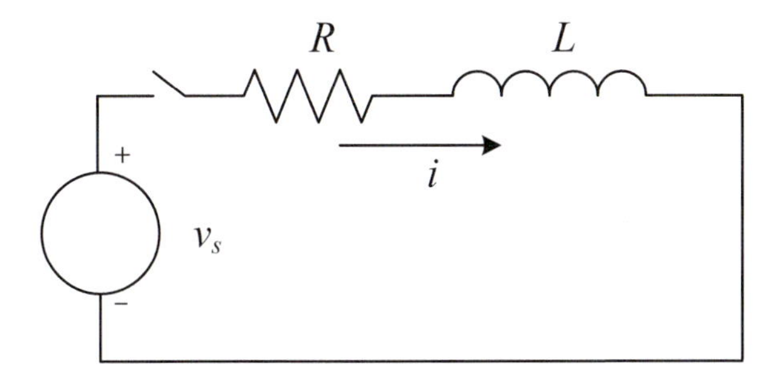

Figure 2.1: An RL circuit.

For a simple system described by (2.1), we can find a closed-form expression for $i(t)$ using the ODE solving techniques from calculus.

The solution $i(t)$ of (2.1) consists of two components $i_1(t)$ and $i_2(t)$:

$$i(t) = i_1(t) + i_2(t), \tag{2.2}$$

where $i_1(t)$ is the forced component or the steady-state response, which is a special solution that satisfies (2.1), and $i_2(t)$ is the transient component which satisfies the homogeneous equation

$$0 = Ri_2(t) + L\frac{di_2(t)}{dt}. \tag{2.3}$$

For $i_1(t)$, we may guess a solution. For example, by comparing the left side of (2.1), V_{DC}, with the right side of the equation, $Ri + \frac{di}{dt}$, we guess that $i_1(t)$ is a constant with its derivative as zero. Then the right side becomes Ri_1. Therefore, we have:

$$i_1 = \frac{V_{DC}}{R}. \tag{2.4}$$

For i_2, the general form for the first-order homogeneous ODE is $Ke^{s(t-t_0)}$, where K and s are constants. Replacing $i_2(t)$ with $Ke^{s(t-t_0)}$ in (2.3), we have:

$$0 = K(R + Ls)e^{s(t-t_0)}. \tag{2.5}$$

To make the right side zero, we must have:

$$s = -\frac{R}{L}.$$

Using the initial condition, we may find K. First, we find the expression for $i(t)$:

$$i(t) = \frac{V_{DC}}{R} + Ke^{-\frac{R}{L}(t-t_0)}. \tag{2.6}$$

At $t = t_0^+$ ($^+$ stands for the moment right after the switch is turned on), we know that current is kept at 0 since the circuit has an inductor and the current through an inductor cannot have a sudden change. Therefore,

$$0 = \frac{V_{DC}}{R} + K \Longrightarrow K = -\frac{V_{DC}}{R}. \tag{2.7}$$

The final expression of $i(t)$ is as follows.

$$i(t) = \frac{V_{DC}}{R}\left(1 - e^{-(t-t_0)/\tau}\right), t \geq t_0^+ \tag{2.8}$$

where $\tau = \frac{L}{R}$.

τ is named as time constant. When $t = t_0 + \tau$,

$$i(t_0 + \tau) = (1 - e^{-1})i(\infty) = 0.63 \times i(\infty),$$

where $i(\infty)$ is the steady-state value of the current and $i(\infty) = \frac{V_{DC}}{R}$. A time constant indicates how fast the system responds. It is an important measure of system dynamics. Another measure is bandwidth, which will be mentioned in Chapter 3. High bandwidth indicates a fast response.

Alternatively, we can obtain the closed-form expression through the Laplace transform. In the Laplace domain, (2.1) is expressed as:

$$\frac{V_{DC}}{s} = RI(s) + L(sI(s) - i(t_0^+)) \tag{2.9}$$

where $I(s) = \mathcal{L}(i(t))$ is the Laplace-domain expression of the current.

At t_0^+, the current is 0. This is due to the fact that initially the circuit is open and there is no current. Further, there is an inductor. Current through an inductor cannot jump even when the switch turns on. Therefore, $i(t_0^+) = 0$.

Based on (2.9), we have the current's expression as:

$$I(s) = \frac{V_{DC}}{s(R + Ls)} \tag{2.10a}$$

$$= \frac{V_{DC}}{R}\left(\frac{1}{s} - \frac{1}{s + R/L}\right) \tag{2.10b}$$

An inverse Laplace transform will lead to the time-domain expression as follows.

$$i(t) = \frac{V_{DC}}{R}\left(1 - e^{-(t-t_0)/\tau}\right), t \geq t_0^+ \tag{2.11}$$

where $\tau = L/R$.

The expressions of $i(t)$ derived based on calculus and Laplace transform are the same.

For a complicated system model, it is not easy to find closed-form time-domain expressions for the desired variables. Instead, numerical integration will be conducted to find the values of ODE variables over time.

In Section 2.2, an overview of numerical integration methods will be introduced. This section is followed by an example on RLC circuit simulation in Section 2.3. In Section 2.4, dynamic model building and simulation in MATLAB/Simulink is explained and demonstrated. In Section 2.5, MATLAB commands that can conduct dynamic simulation for linear time-invariant (LTI) systems are given.

2.2 Numerical integration methods

Numerical integration methods are usually covered in textbooks on numerical methods, e.g., Crow (2015); Chapra and Canale (2012). Chapter 5 of Crow (2015) gives a detailed treatment on numerical integration methods, including accuracy analysis and numerical stability analysis. This textbook focuses on applications and hands-on training. Numerical integration methods will be introduced along with codes and MATLAB® commands. Three methods will be discussed in this section: the Forward–Euler method, Runge–Kutta method, and Trapezoidal method.

2.2.1 Forward–Euler method

Give a set of differential equations

$$\frac{dx}{dt} = f(x), \tag{2.12}$$

where $x \in \mathbb{R}^n$ and f is a mapping $f : \mathbb{R}^n \Rightarrow \mathbb{R}^n$. The objective of numerical integration is to find $x(t_1), x(t_2), \cdots, x(T)$ in the time interval of $t_0 \sim T$ for a given initial condition $x(t_0)$.

If we select a constant step size h, then

$$\begin{aligned} t_1 &= t_0 + h \\ \vdots &= \vdots \\ t_k &= t_{k-1} + h = t_0 + kh \\ \vdots &= \vdots \\ t_N &= t_{N-1} + h = t_0 + Nh \end{aligned}$$

We aim to find $x_k, k = 1, \cdots, N$, where $N = (T - t_0)/h$, and x_k is the approximated value of $x(t_k)$.

Based on Taylor's series, $x(t_{k+1})$ can be expressed by $x(t_k)$ and the derivatives of x evaluated at t_k.

$$\begin{aligned} x(t_{k+1}) &= x(t_k) + \dot{x}(t_k)h + O(h^2) \\ &\approx x(t_k) + \dot{x}(t_k)h \\ &= x(t_k) + f(x(t_k))h \end{aligned}$$

where $O(\cdot)$ stands for high-order terms.

Replacing $x(t_k)$ by its approximated value x_k, we have:

$$x_{k+1} = x_k + f(x_k)h. \tag{2.13}$$

(2.13) is the Forward Euler method. The accuracy of the method is evaluated by $x(t_{k+1}) - x_{k+1}$. If we assume that the error between $x(t_k)$ and x_k can be ignored, we can find that

$$x(t_{k+1}) - x_{k+1} = O(h^2) \tag{2.14}$$

Therefore, the Euler method has an accuracy of $O(h^2)$.

For the RL circuit example, we can quickly write a Python code to conduct the simulation.

```
import math,pylab
# define the dynamic equation to compute di/dt # R*i + L. di/dt = v
def fun_didt(R, L, v, i):
    di_dt = (v-R*i)/L;
    return di_dt

# Use trapezoidal method to conduct numerical integration
# step 1, initial condition
# voltage is a dc voltage. for all time being, voltage is 5V.
step_size = 0.01
n_steps = 1000
v = 5.0
R = 0.1
L = 0.1
i_data =[]
v_data =[]
i = 0
for k in range(n_steps):
    v_data.append(5.0)
    i_data.append(i)
    # compute current
    i = i + fun_didt(R, L, v, i)*step_size;
tt = [k*step_size for k in range(n_steps)]
pylab.plot(tt, i_data)
pylab.xlabel('time (sec)');
pylab.show()
```

First, a function *fun_didt* is defined. This function computes the derivative of current for given source voltage, current and known parameters (R and L). The simulation step h is 0.01 seconds and total 1000 steps will be computed using numerical integration. The DC voltage is 5 V. The resistance is 0.1 Ω. The inductance is 0.1 H. Two lists *i_data* and *v_data* are used to store current and voltage computed at every step. Initial current is set to be 0.

The iteration is carried out for 1000 steps and at every step, the computed current and voltage are appended to the two lists, respectively. At every step, *fun_didt* is called and Forward–Euler is carried out to compute the current at the next step.

Finally, pylab is used to make plots. The dynamic simulation result from the above Python code is shown in Figure 2.2.

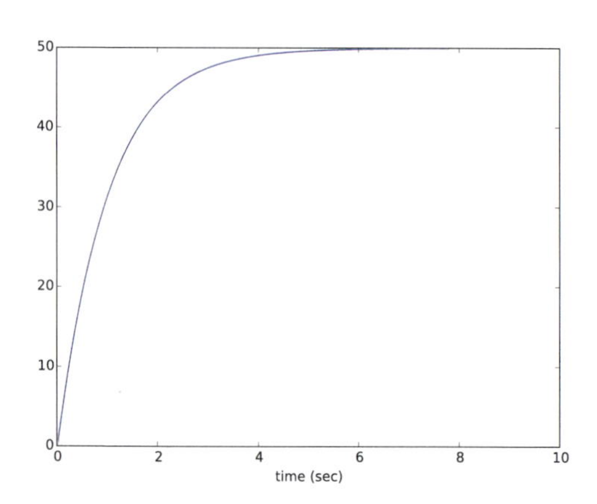

Figure 2.2: Time-domain response of the circuit current using Forward–Euler method.

In MATLAB, ode1 is the integration solver using the Euler method.

2.2.2 Runge–Kutta method

The Euler method is not popular due to its low accuracy. In the Runge–Kutta method, the second-order term is preserved and approximated using numerical computation. The following fourth-order Runge–Kutta has the accuracy of $O(h^5)$. The MATLAB ODE solver for Runge–Kutta is ode4.

$$x_{k+1} = \frac{1}{6}\left(k_1 + 2k_2 + 2k_3 + k_4\right) \tag{2.15a}$$

$$k_1 = f(x_k) \tag{2.15b}$$

$$k_2 = f(x_k + \frac{h}{2}k_1) \tag{2.15c}$$

$$k_3 = f(x_k + \frac{h}{2}k_2) \tag{2.15d}$$

$$k_4 = f(x_k + hk_3) \tag{2.15e}$$

2.2.3 Trapezoidal method

The Trapezoidal method has the accuracy of $O(h^3)$ and has been adopted in Power System Toolbox Chow and Cheung (1992), a free MATLAB toolbox for power system dynamic simulation developed by J. Chow in the early 1990s.

In the Forward–Euler method, the derivative of x at the period of $[t_k, t_{k+1}]$ is assumed to be $f(x(t_k))$. In the Trapezoidal method, the derivative $f(x)$ is approximated by a line connecting $f(x(t_k))$ and $f(x(t_{k+1}))$ as shown in Figure 2.3. Therefore, we will use the trapezoidal area to replace the integration of $f(x)$ from t_k to t_{k+1}.

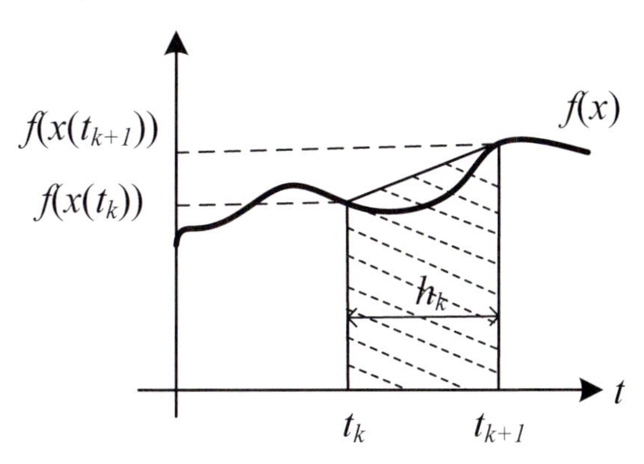

Figure 2.3: Trapezoidal method.

$$\begin{aligned} x_{k+1} &= x_k + \int_{t_k}^{t_{k+1}} f(x)dt \\ &\approx x_k + \frac{h}{2}\left(f(x_k) + f(x_{k+1})\right) \end{aligned} \tag{2.16}$$

Note the Trapezoidal method requires the state derivative $f(x)$ be evaluated at t_{k+1}. The Forward–Euler method is first applied to give an estimation of x_{k+1}, notated as $\tilde{x}_{k+1}$, as shown in (2.17a). Afterwards, the Trapezoidal method is applied to find the state at step $k+1$ as shown in (2.17b).

$$\tilde{x}_{k+1} = x_k + hf(x_k) \tag{2.17a}$$

$$x_{k+1} = x_k + \frac{h}{2}(f(x_k) + f(\tilde{x}_{k+1})) \tag{2.17b}$$

Table 2.1 gives a comparison of the three methods for numerical integration results of the RL circuit. The values presented are $x_k - x(t_k)$, where

$x(t_k)$ is computed using the analytical close-form in (2.8). It can be seen that the 4th order Runge–Kutta method (RK4) gives the most accurate result, while the Trapezoidal method's accuracy is much higher than that of the Forward–Euler method.

Table 2.1: Comparison of the errors of the three methods

Time	Euler	RK4	Trapezoidal
0.01	−2.49e−03	4.16e−11	8.31e−06
0.02	−4.93e−03	8.24e−11	2.88e−05
0.03	−7.33e−03	1.22e−10	6.13e−05
0.04	−9.67e−03	1.61e−10	1.00e−04
0.05	−1.20e−02	1.99e−10	1.60e−04
0.06	−1.42e−02	2.37e−10	2.20e−04
0.07	−1.64e−02	2.74e−10	3.04e−04
0.08	−1.85e−02	3.10e−10	3.92e−04
0.09	−2.07e−02	3.46e−10	4.89e−04

Table 2.1 was generated by the following Python code.

```
import math,pylab

# define the dynamic equation to compute di/dt
# R*i + L. di/dt = v
def fun_didt(R, L, v, i):
    di_dt = (v-R*i)/L;
    return di_dt

step_size = 0.01; n_steps = 10;
v = 5.0; R = 0.1; L = 0.1;
i_data =[]; i_Eu = [];i_RK = []; i_Tr = []; v_data =[];
i1 = 0; i2 = 0; i3 = 0;

for k in range(n_steps):
    v_data.append(5.0)
    i_data.append(5.0/R*(1-math.exp(-R*step_size*k/L)))
    i_Eu.append(i1)
    i_RK.append(i2)
    i_Tr.append(i3)
    # Euler
    i1 = i1 + fun_didt(R, L, v, i1)*step_size;
    # Trapzoidal
    i3 = i3 + 0.5*(fun_didt(R, L, v, i1)+ fun_didt(R, L, v, i3))*step_size;
    # RK4
    k1 = fun_didt(R, L, v, i2)
    k2 = fun_didt(R, L, v, i2 + 0.5*step_size*k1)
    k3 = fun_didt(R, L, v, i2 + 0.5*step_size*k2)
```

```
    k4 = fun_didt(R, L, v, i2 + step_size*k3)
    i2 = i2 + step_size/6*(k1+2*k2+2*k3+k4)

# print a table
for k in range(n_steps):
    print step_size*k,i_data[k]-i_Eu[k],i_data[k]-i_RK[k],i_data[k]-i_Tr[k]
```

2.3 Dynamic simulation for an RLC circuit

In this section, dynamic simulation is carried out using Python programming and MATLAB's numerical integration solver such as ode45.

The RLC circuit is shown in Figure 2.4. The state-space model of the

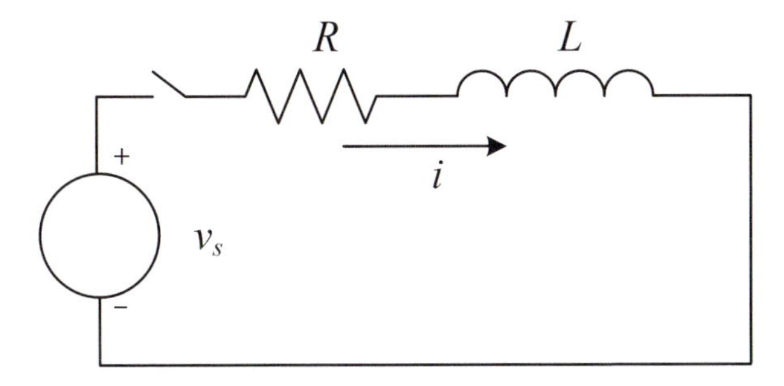

Figure 2.4: An RLC circuit.

RLC circuit is first derived. We are only interested in differential equations. Therefore, no integral should appear in the model. For the RLC circuit, the derivative of the line current i is proportionally related to the voltage drop across the inductor v_L; the derivative of the capacitor voltage v_C is proportionally related to the current through the capacitor i. Therefore, we can write the two first-order differential equations and further apply Kirchhoff's voltage law (KVL) to write the loop voltage equation.

$$L\frac{di}{dt} = v_L \tag{2.18a}$$

$$C\frac{dv_C}{dt} = i \tag{2.18b}$$

$$-v_s + Ri + v_L + v_C = 0 \tag{2.18c}$$

The two variables i and v_C are called state variables. The above model is termed as dynamic algebraic equations (DAEs). We will get rid of the algebraic equation and have a set of ODEs.

$$L\frac{di}{dt} = v_s - Ri - v_C \tag{2.19a}$$

$$C\frac{dv_C}{dt} = i \tag{2.19b}$$

Note that in the above equations, the derivatives of the state variables are expressed by themselves along with a given input v_s.

A general expression for the above system is as follows.

$$\dot{x} = f(x, u) \tag{2.20}$$

where u is input and x is the vector of the state variables.

For the RLC circuit, $x = [i, v_C]^T$ and $u = v_s$. This is a linear system and we can write the following linear state-space model.

$$\dot{x} = \underbrace{\begin{bmatrix} \frac{-R}{L} & \frac{-1}{L} \\ \frac{1}{C} & 0 \end{bmatrix}}_{A} x + \underbrace{\begin{bmatrix} \frac{1}{L} \\ 0 \end{bmatrix}}_{B} \underbrace{v_s}_{u} \tag{2.21}$$

Further let us define the output of the model y as the same as x. Then we have this equation:

$$y = Cx + Du \tag{2.22}$$

where $C = I$ is an identity matrix and $D = [0, 0]^T$.

If we are only interested in having the current as the output, then

$$y = \begin{bmatrix} 1 & 0 \end{bmatrix} x \tag{2.23}$$

For a nonlinear continuous system with state variables constant at a steady-state, a linear model can be derived by evaluating the model at a steady-state operating condition x_0 and u_0 ($f(x_0, u_0) = 0$) using small perturbation.

$$\begin{aligned} \Delta\dot{x} &= \frac{d(x_0 + \Delta x)}{dt} = f(x_0 + \Delta x, u_0 + \Delta u) \\ &\approx f(x_0, u_0) + \left.\frac{\partial f}{\partial x}\right|_{x_0,u_0} \Delta x + \left.\frac{\partial f}{\partial u}\right|_{x_0,u_0} \Delta u \\ &= \left.\frac{\partial f}{\partial x}\right|_{x_0,u_0} \Delta x + \left.\frac{\partial f}{\partial u}\right|_{x_0,u_0} \Delta u \end{aligned} \tag{2.24}$$

2.3.1 Trapezoidal method-based simulation in Python

The following Python code is developed to simulate the RLC circuit. Note that a function to compute the derivatives dx/dt is first defined. The inputs to the function include the RLC parameters, the state variable vector x and the source voltage v_s. The main routine carries out trapezoidal numerical integration. At each step k, the derivatives dx/dt are computed by calling the function. Based on the derivative, the next step's state is estimated using Forward–Euler method. Based on this estimation, the derivative at step $k+1$ is computed. Using the two derivatives, the $k+1$ step state x_{k+1} is then computed.

```
import math,pylab
# define the dynamic equation to compute di/dt
# R*i + L. di/dt +vc = v # c
dvc/dt = i def fun_dxdt(R, L, C, vs, x):
    dx_dt = [-R/L*x[0]-x[1]/L+vs/L, x[0]/C]
    return dx_dt

# Trapezoidal method to conduct numerical integration
# step 1, initial condition
# voltage is a dc voltage. for all time being, voltage is 1V.
step_size = 0.001
n_steps = 1000
v = 1
R = 0.1
L = 0.01
C =0.001
x_data =[]
v_data =[]
x = [0, 0]
x1 =[0, 0]
x2 =[0, 0]
for k in range(n_steps):
    v_data.append(v)
    x_data.append(x[:])
    x = x2
    # compute current
    dx_dt = fun_dxdt(R, L, C, v, x)
    for i in range(2):
        x1[i] = x[i] + dx_dt[i]*step_size
    dx_dt_est = fun_didt(R, L, C, v, x1)
    for i in range(2):
        x2[i]  = x[i] + 0.5*(dx_dt[i]+dx_dt_est[i])*step_size
tt = [k*step_size for k in range(n_steps)]
pylab.plot(tt, x_data)
pylab.xlabel('time (sec)');
```

```
pylab.show()
```

Figure 2.5 shows the simulation results from the Python code.

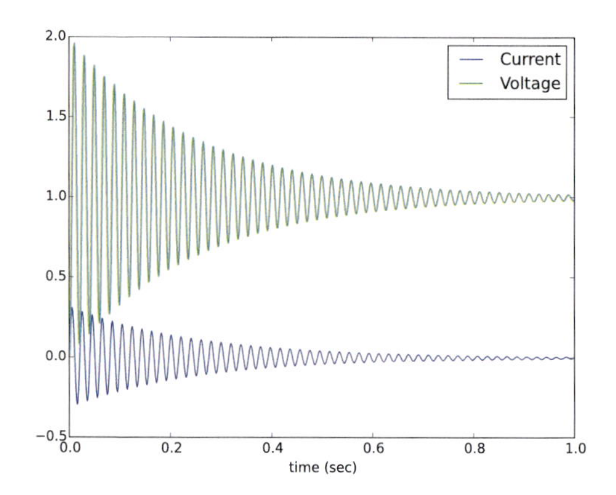

Figure 2.5: RLC circuit simulation results.

2.3.2 MATLAB® ODE solver-based dynamic simulation

MATLAB's ODE solvers provide a convenient approach to conduct numerical integration. We only need to define $f(x,u)$ and then call the solver.

In an m-file named fun_RLC.m, define a function as fun_RLC to compute $f(x,u)$. To use ODE solvers, we first define u. Here we set $u = 1$ to initiate a step response.

```
function xdot = fun_RLC(t,x,par)
xdot = [ -par.R/par.L, -1/par.L;1/par.C, 0]*x + [1/par.L; 0];
```

The main file calls the ODE solver.

```
par.R = 0.1; par.L = 0.01; par.C = 0.001;
[t,y]=ode45(@(t,x) fun_RLC(t,x,par),[0 1],[0 0]);
plot(t,y);
```

Note that the ODE solvers require the function name that computes the state vector derivative. The second and third inputs of the ode45 function are time span and initial state variables. The simulation results obtained are the same as those shown in Figure 2.5.

2.4 MATLAB/Simulink-based dynamic simulation

MATLAB/Simulink offers a graphic user interface (GUI) for model building. Once a model is built, numerical integration can be carried out automatically after clicking a run button. In Simulink, we do not need to write the steps of the Trapezoidal method as presented in section 2.3.1. Nor do we need to explicitly call the ODE solvers. ODE solvers are specified in the configuration dialog box in Simulink setup. We present two types of dynamic model building techniques. The first is based on an integrator. This technique is more aligned to our understanding of differential equations where derivatives and states are separated by an integrator. The alternative approach is S-function. In an S-function block, we aggregate the entire system into one block with inputs and outputs.

2.4.1 Integrator-based model building

The critical step of the model building procedure is to identify the state variable vector x and compute its derivative $\dot{x} = f(x, u)$ from the input variable vector u and itself x. Then the state and its derivative will be linked by an integrator. With initial state $x(0)$ set in the integrator, this model is ready to run.

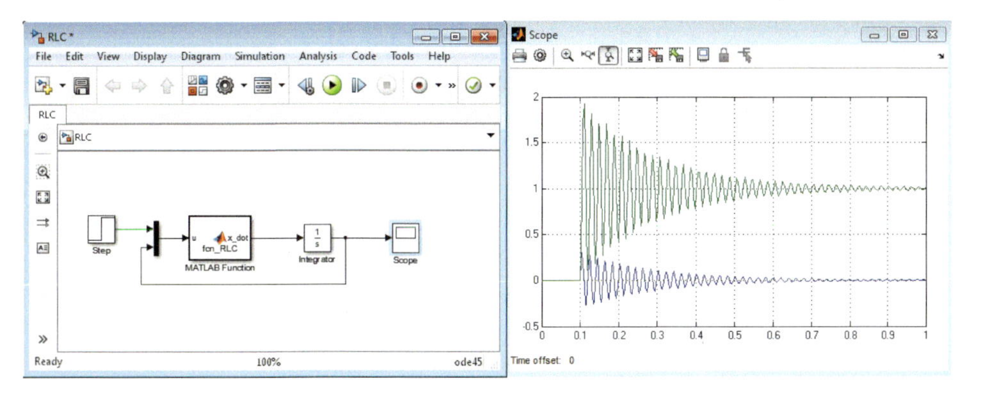

Figure 2.6: Integrator-based dynamic model building in Simulink.

Figure 2.6 shows the building blocks for a dynamic model that can be expressed as $\dot{x} = f(x, u)$. Note that an integrator is used. Input of the integrator should be $f(x, u)$ while the output of the integrator is x. $f(x, u)$ is computed using a MATLAB embedded function. The input of the function is a vector consisting of the voltage source v_s and the state variables x. v_s and x are concatenated to generate a single vector. Inside the function fcn_RLC, v_s and x are separated. Note that the function has another input *par*, which is a structure that contains the values of R, L, and C.

```
function x_dot = fcn_RLC(u, par)
%#codegen
vs = u(1);
 x = u(2:end);
 A = [ -par.R/par.L, -1/par.L;
       1/par.C, 0];
 B = [ 1/par.L; 0];
 x_dot = A*x+B*vs;
 return
```

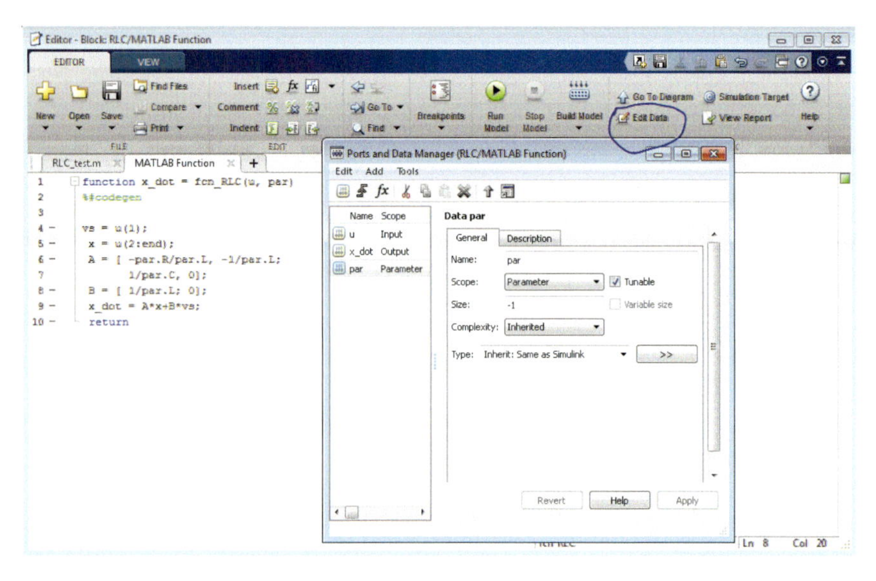

Figure 2.7: Set an input of the MATLAB embedded function as a parameter.

As shown in Figure 2.7, MATLAB treats *par* as another input. In order to treat *par* as a given parameter, we need to edit the data and set *par* as a parameter.

One more important thing is to set the initial values of the integrator correctly, especially the size of $x(0)$. In this example, $x(0)$ should be a vector consisting of the initial current value and the initial voltage value. Therefore the initial vector is set as $[0; 0]$ or zeros(2,1) as shown in Figure 2.8.

In the simulation stage, we can configure MATLAB/Simulink's configuration dialog box to set the simulation period, numerical integration method, and step size.

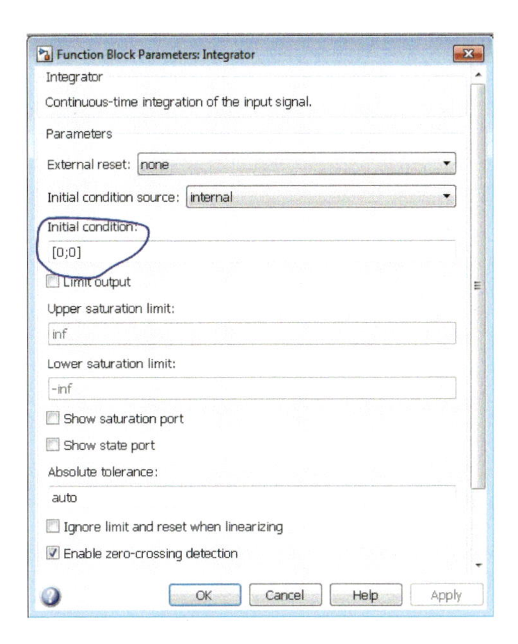

Figure 2.8: Set $x(0)$ appropriately.

2.4.2 S-function-based model building

In the S-function approach, the entire system is defined in one block as shown in Figure 2.9. An S-function has a fixed template to define the dimension of the system, initial state, state derivative, etc. For this RLC example, we use a continuous system template.

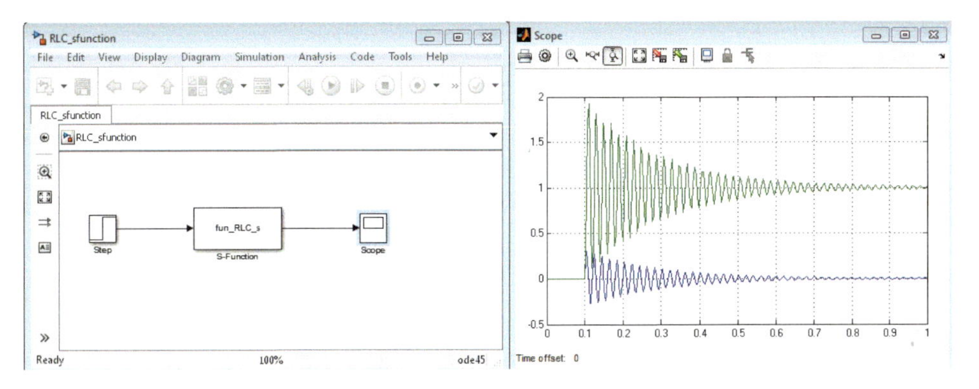

Figure 2.9: Model building in Simulink using S-function.

The S-function has three scenarios to consider: (i) initialization; (ii) derivative calculation; and (iii) output definition. The initial state vector setting, dimensions of the input, output and state, are all set in the first task, which is realized by a function *mdlInitializeSizes*. The second task, derivative calculation, is realized by the function *mdlDerivatives*. The third task, output definition, is realized using another function *mdlOutputs*. The code of the S-function is shown as follows.

```
function [sys,x0,str,ts,simStateCompliance] = fun_RLC_s(t,x,u,flag)
switch flag,
  % Initialization %
  case 0,
    [sys,x0,str,ts,simStateCompliance]=mdlInitializeSizes;
  % Derivatives %
  case 1,
    sys=mdlDerivatives(t,x,u);
  % Outputs %
  case 3,
    sys=mdlOutputs(t,x,u);
 % other cases related to discrete systems
  case {2,4,9}
    sys=[];
  otherwise
    DAStudio.error('Simulink:blocks:unhandledFlag', num2str(flag));
end

function [sys,x0,str,ts,simStateCompliance]=mdlInitializeSizes
sizes = simsizes;
sizes.NumContStates  = 2;   % x is a vector of 2.
sizes.NumDiscStates  = 0;
sizes.NumOutputs     = 2;
sizes.NumInputs      = 1;
sizes.DirFeedthrough = 0;
sizes.NumSampleTimes = 1;   % at least one sample time is needed
sys = simsizes(sizes);
% initialize the initial conditions
x0  = [0;0];
str = [];
% initialize the array of sample times%
ts  = [0 0];
simStateCompliance = 'UnknownSimState';

% case 1: derivative
function sys=mdlDerivatives(t,x,u)

par.R = 0.1;
par.L = 0.01;
par.C = 0.001;
 A = [ -par.R/par.L, -1/par.L;
       1/par.C, 0];
```

```
 B = [ 1/par.L; 0];
 x_dot = A*x+B*u;
sys = x_dot;

% case 3: output
function sys=mdlOutputs(t,x,u)
sys = x;
```

2.5 MATLAB commands for linear system simulation

The simulation methods discussed in Section 2.4 apply to a general dynamic system. For linear systems, we may use MATLAB commands to conduct simulation directly. MATLAB's control toolbox offers many functions related to linear time invariant (LTI) system analysis and simulation. A few lines of commands can help us build an LTI model and conduct time-domain simulations. This saves us from the coding tasks.

Using the RLC circuit example, we will build an LTI model using the transfer function or state-space.

2.5.1 Define the linear models

The first step is to define the LTI model of the RLC circuit. The model has been derived in Section 2.2 and is in the form of the following equation.

$$\begin{aligned} \dot{x} &= Ax + Bu \\ y &= Cx + Du \end{aligned} \tag{2.25}$$

This state-space model can be defined in MATLAB by one line if A, B, C, and D have been defined.

```
sys1 = ss(A,B,C,D)
```

The transfer function of the model can also be found.

```
sys2 = tf(sys1)
```

Sys2 will be in the form of a Laplace transfer function.

Another approach is to define the model in the frequency domain. Let us examine the impedance model of each element. Impedance model is a term used to describe the ratio of incremental voltage and incremental current in the frequency domain $\frac{\Delta V(s)}{\Delta I(s)}$. Impedance modeling technique is a widely used modeling technique in power electronic converter analysis. For R, the impedance model is R. For an inductor, the impedance model is Ls and for a capacitor, the impedance model is $\frac{1}{Cs}$.

Therefore, we can find the incremental current expression by Ohm's law. The transfer function from $\Delta V(s)$ to $\Delta I(s)$ is as follows.

$$\frac{\Delta I(s)}{\Delta V(s)} = \frac{1}{R + Ls + 1/(Cs)} \tag{2.26}$$

For a linear system, we usually ignore Δ when expressing the transfer function. Further, we may ignore (s).

$$\frac{I}{V_s} = \frac{1}{R + Ls + 1/(Cs)} \tag{2.27a}$$

$$\frac{V_C}{V_s} = \frac{I \times \frac{1}{Cs}}{V_s} \tag{2.27b}$$

$$\frac{V_L}{V_s} = sL\frac{I}{V_s} \tag{2.27c}$$

The above transfer functions define the input/output relationships of the current, the capacitor voltage, and the inductor voltage versus the input voltage. They can be expressed using the Laplace operator s in MATLAB.

```
s = tf('s');
sys2 = 1/(R+ L*s + 1/(C*s));
sys3 = sys2 *1/(C*s);
sys4 = L*s*sys2;
```

The second line defines the input/out relationship of the source voltage and the current. The third line defines the input/output relationship of the source voltage and the capacitor voltage. The fourth line defines the input/output relationship between the source voltage and the inductor voltage.

2.5.2 Time-domain responses

With the model defined, we can carry out time-domain simulation using MATLAB commands.

Example 1: please give the time-domain response of the current $i(t)$ when the source voltage is subject to a step response.

Solution: Step response can be found using the MATLAB command *step*. The codes are shown below.

```
R = 0.1; L = 0.01; C = 0.001;
s = tf('s');
sys2 = 1/(R+ L*s + 1/(C*s));
step(sys2);
```

Example 2: please give the time-domain response of the current $i(t)$ when the source voltage changes from 0 to a sinusoidal voltage with 6 Hz frequency.

Solution: Generally, we can use *lsim* to conduct simulation for any given input u. The codes are shown below.

```
T = 0:0.001:1.2;
u = cos(37.7*T);
lsim(sys2, u, T);
```

Figure 2.10 shows the plots given by the two commands *step* and *lsim*.

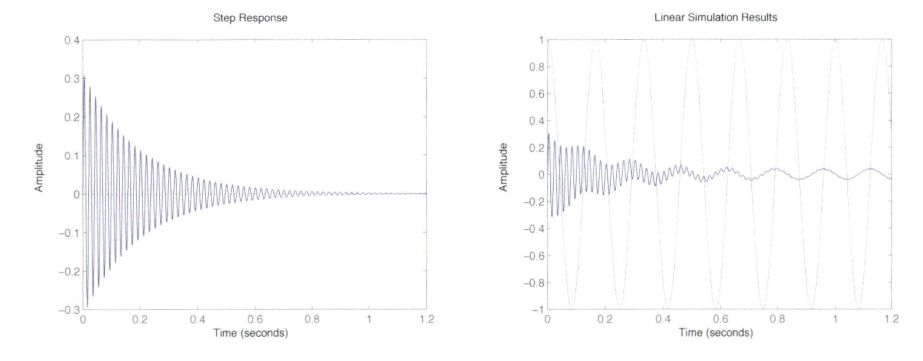

Figure 2.10: Step response and sinusoidal input response.

2.5.3 Linear system analysis

With a single-input-single-output (SISO) system, we can also carry out linear system analysis using the MATLAB command *bode* to obtain Bode plots, *pole* to find the system poles, *zero* to find the system zeros, and *pzmap* to find the poles and zeros in the real-imaginary space.

Figure 2.11 was generated by the following code.

```
R1 = 0.1; R2 = 0.001;R3 = 1;
  L = 0.01;C = 0.001;
s = tf('s');
% current
```

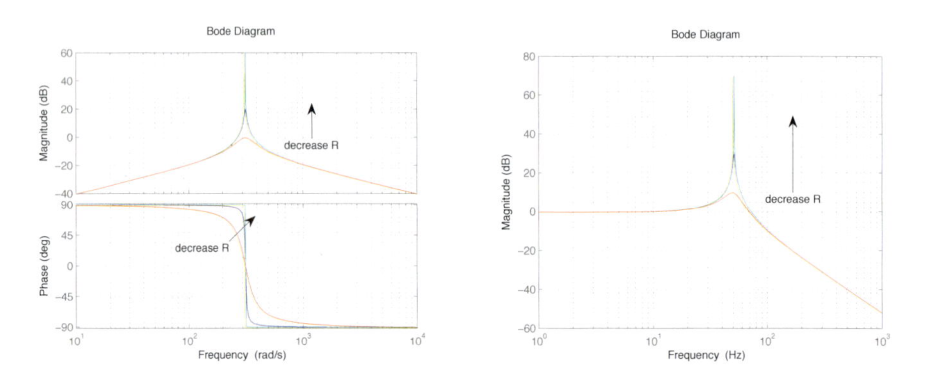

Figure 2.11: Bode plots of I/V_s and V_c/V_s.

```
G1 = 1/(R1+ L*s + 1/(C*s));
G2 = 1/(R2+ L*s + 1/(C*s));
G3 = 1/(R3+ L*s + 1/(C*s));

figure(1);
bode(G1,G2,G3);
grid;

% capacitor voltage
figure(2);
h = bodeplot(G1/(C*s), G2/(C*s), G3/(C*s));
grid;
 % Change units to Hz and make phase plot invisible
setoptions(h,'FreqUnits','Hz','PhaseVisible','off');
```

The Bode plots of the magnitude in Figure 2.11 show that there is a peak at 500 Hz. This indicates that there are oscillations with 500 Hz. With a smaller resistance, the peak is more obvious. A higher peak in the Bode plot indicates less damping of the oscillation at 500 Hz. Examining the time-domain simulation plots in Figure 2.5 and Figure 2.10, we can find that the frequency of the oscillations is indeed about 500 Hz.

2.6 Summary

In this chapter, dynamic simulation, the main tool for control validation, is introduced. Throughout this chapter, an RLC example is demonstrated for simulation in Python and MATLAB/Simulink. Further, linear system dynamic simulation commands in MATLAB are introduced. With linear systems, analysis (e.g., Bode plots) can be conducted. The indication from linear system analysis can be verified by the time-domain simulation results.

Throughout this textbook, linear system analysis is the main analytical tool that will be used for control design and stability analysis, while dynamic simulation is the validation tool that will be used for both linear and nonlinear systems.

Exercises

1. A voltage source is serving an RLC series connected circuit. Let $R = 0.01\Omega$, $L = 0.01$ H, $C = 0.001$ F. The compensation degree of the system is X_c/X_L, approximately 70.36%. Find its current response for a step response of the voltage source, and a sinusoidal 60 Hz input (amplitude 1 V) of the voltage source. Use Laplace transformation to find the current in the Laplace domain and current in the time domain.

2. Use MATLAB linear system analysis tools to define a linear system for the above RLC circuit. Treat the voltage source as the input while the current is the output. Give a set of Bode plots of the system by varying R. Notate the plot properly. Use MATLAB function *step* to examine the dynamic response of the current with a step response of the voltage source. Use MATLAB function *lsim* to examine the dynamic response of the current with a sinusoidal input.

3. For the above RLC circuit, build a two-order state-space model. The state variables are the current and the voltage across the capacitor. Use MATLAB function *ode* to simulate the dynamic response of the current for a step response and a sinusoidal input.

Chapter 3

Frequency Control

3.1 Important facts

Before we dive into the topic of frequency control, several facts are stated as follows.

- For synchronous machines, **frequency *is* speed**. The values of electricity frequency and the speed are the same in per unit. If a machine is running at the nominal speed, say 3600 revolutions per minute (rpm) for a 2-pole machine, the corresponding stator electricity frequency is 60 Hz. This mechanism is determined by Faraday's law (change of the flux linkage linked to a circuit induces voltage or electromotive force (EMF): $\frac{d\lambda}{dt} = v$). If the flux linkage linked to the stator circuits is sinusoidal and has a frequency ω, then the voltage induced will be sinusoidal and has a frequency ω.

  When the rotor of an ac machine is rotating at 3600 rpm, and the rotor circuit has a constant excitation current i_F, the effect is to set up a rotating magnetic field (flux) with a rotating speed at 3600 rmp. This rotating magnetic field will cause a stator circuit to encompass a flux linkage in the form of $\hat{\lambda}\cos(2\pi 60t + \theta_0)$, where θ_0 is the initial rotor position. In turn, the induced voltage in the stator circuit will have a frequency at 60 Hz.

  The details will be covered in Chapter 4 Synchronous Generator Models where electromagnetic fields will be examined.

- The second fact is **frequencies at different locations in an electric system (all components are connected) are the same** at steady-state. If there is a load change, generators will react towards the change by adjusting speeds. In the end, all speeds should be the same in per unit. This is due to the electric system's interconnected condition. As long as this system is working, voltage phasors everywhere are related. For any two buses i and j, the two voltages $v_i(t)$ and $v_j(t)$ should have the same frequency and their angles may be different but their difference is kept constant at steady-state, i.e., $v_i(t) = V_i \cos(\omega_0 t + \theta_i)$, $v_j(t) = \cos(\omega_0 t + \theta_j)$, and $\theta_i - \theta_j = \text{constant}$. θ_i and θ_j may be time varying.

This chapter deals with frequency control of power systems with synchronous generators as the main components. For any control problem, the design procedure consists of understanding of the control objectives, figuring out the plant model, the inputs and outputs of the controller, the controller design, and finally validation of the controller's performance.

Therefore, this chapter is organized to fit the control design procedure. Section 3.2 presents the plant model. Section 3.3 presents the first control objective, steady-state frequency deviation reduction, and the design methods. Section 3.4 presents the second control objective, steady-state frequency deviation elimination, and the design methods. Automatic generation control (AGC) is covered in this section. Section 3.5 presents validation results using time-domain simulation. Section 3.6 provides more examples related to frequency control analysis. MATLAB codes are provided to help readers improve their skills of conducting linear system analysis.

3.2 Plant model: Swing equations

In this section, the plant model for frequency control will be examined. We will start from Newton's second law for a rotating mass to derive the swing equation for a synchronous generator. Assumptions are then made to simplify the equation. Further, linearized equation for small-signal perturbations are derived.

With the swing equation given, we then proceed to examine two systems for their frequency responses when they are subject to a load change or mechanical power change.

3.2.1 Newton's Law for a rotating mass

The swing equation comes from Newton's second law for a rotating mass. In the case of a synchronous generator, the rotating mass refers to the rotor. Newton's Law states that the acceleration speed is proportional to the net torque:

$$J\frac{d\omega}{dt} = T_m - T_e \tag{3.1}$$

where J is the inertia of the rotor with units as kilogram meter squared (kg.m^2) or Joules second^2 (J.s^2). ω is the rotating speed in rad/s, and torques are in standard unit Newton.meter (N.m). T_m is the mechanical torque generated by the prime mover, while T_e is the electromagnetic torque generated by the electromagnetic field.

For power system engineers, power is used more often than torque. Therefore, the above equation will be written in terms of the mechanical power P_m and the electric power generated by the electromagnetic field P_e.

Note that $P_m = T_m\omega_m$ where ω_m is the mechanical speed, and $P_e = T_e\frac{2}{P}\omega_e$, where ω_e is the electricity frequency in the stator circuit and P is the poles of the machine. For two-pole machines, $P_e = T_e\omega_e$.

For a two-pole synchronous generator, the mechanical speed ω_m and the stator's electricity frequency ω_e are the same. Therefore, we use ω to represent both the rotating speed and the electricity frequency. The swing equation in (3.1) becomes:

$$J\frac{d\omega}{dt} = \frac{P_m}{\omega} - \frac{P_e}{\omega}. \tag{3.2}$$

Note that this equation is applicable for synchronous machines only. In the case of induction machines, $T_m = P_m/\omega_m$, and $T_e = P_e/\omega_e$, where ω_m is the rotating speed and ω_e is the electric frequency in the stator circuits. For induction machines, the rotating speed and the electric frequency are not equal. The electric frequency ω_e equals the rotating speed of the rotating magnetic field of a 2-pole machine. The speed of the field is the sum of the mechanical speed ω_m and the rotor circuit current frequency ω_r: $\omega_e = \omega_m + \omega_r$.

If we consider friction in the mechanical system and the friction torque is proportional to the speed, then the above equations will be modified as follows.

$$J\frac{d\omega}{dt} = T_m - T_e - k\omega \tag{3.3}$$

$$J\omega\frac{d\omega}{dt} = P_m - P_e - k\omega^2 \tag{3.4}$$

where k is the coefficient related to friction.

(3.4) is nonlinear in terms of ω. Both $J\omega\frac{d\omega}{dt}$ and $k\omega^2$ are nonlinear. Taylor's expansion is frequently used to obtain a linear expression. In order to obtain a linear expression, an initial condition or a steady-state operating condition should be assumed. In dynamics and control, the initial steady-state condition and final steady-state condition after transients are all termed as equilibrium points.

3.2.2 Swing equation at near nominal speed

If the generator is working at nominal condition with a speed ω_0, we can linearize the above equation through Taylor's expansion evaluated at the nominal condition. The linearized model is applicable for small-signal dynamics at near nominal conditions.

$$\omega\frac{d\omega}{dt} = (\omega_0 + \Delta\omega)\left(\overbrace{\frac{d\omega_0}{dt}}^{0} + \frac{d\Delta\omega}{dt}\right) = \omega_0\frac{d\Delta\omega}{dt} + \Delta\omega\frac{d\Delta\omega}{dt} \tag{3.5a}$$

$$\approx \omega_0\frac{d\Delta\omega}{dt} \tag{3.5b}$$

$$= \omega_0\frac{d\omega}{dt} \tag{3.5c}$$

Note that $\frac{d\omega_0}{dt} = 0$ since ω_0 is constant, $\Delta\omega\frac{d\Delta\omega}{dt}$ contains the product of two small deviations and will be ignored.

For the term $k\omega^2$, linearization is carried out by the general linearizing procedure. For a function $f(x)$, its small deviation evaluated at x_0 is

$$\Delta f \approx \left.\frac{\partial f}{\partial x}\right|_{x_0} \Delta x. \tag{3.6}$$

Therefore

$$k\omega^2 = k\omega_0^2 + \Delta(k\omega^2) \approx k\omega_0^2 + 2k\omega_0\Delta\omega. \tag{3.7}$$

where $\Delta\omega = \omega - \omega_0$.

The Newton's law applicable for conditions near nominal operating point is now represented by

$$J\omega_0\frac{d\omega}{dt} = \widetilde{P_m} - P_e - 2k\omega_0\Delta\omega \tag{3.8}$$

where $\widetilde{P_m} = P_m - k\omega_0^2$.

3.2.3 Swing equation in per unit

The above equation uses physical units. For example, ω is in rad/s, power is in Watts. For power system engineers, power is preferred to be expressed in per unit value. Therefore, both the left-hand side (LHS) and the right-hand side (RHS) of (3.8) will be divided by the power base of the system S_b. This yields

$$\frac{J\omega_0}{S_b}\frac{d\omega}{dt} = \widetilde{P_m^{pu}} - P_e^{pu} - \frac{2k\omega_0}{S_b}\Delta\omega. \tag{3.9}$$

Further, if we use ω^{pu} ($\omega^{pu} = \frac{\omega}{\omega_0}$), then the above equation becomes

$$\frac{J\omega_0^2}{S_b}\frac{d\omega^{pu}}{dt} = \widetilde{P_m^{pu}} - P_e^{pu} - \frac{2k\omega_0^2}{S_b}\Delta\omega^{pu} \tag{3.10}$$

by replacing $\omega = \omega_0\omega^{pu}$ and $\Delta\omega = \omega_0\Delta\omega^{pu}$.

Define

$$H \triangleq \frac{J\omega_0^2}{2S_b}. \tag{3.11}$$

H is the ratio of the kinetic energy at the nominal speed of the rotor versus the power base. H has a unit in seconds.

Hereafter, we will ignore the superscript pu. The Newton's law becomes:

$$2H\frac{d\omega}{dt} = \widetilde{P_m} - P_e - D_1\Delta\omega. \tag{3.12}$$

where $D_1 = \frac{2k\omega_0^2}{S_b}$, ω, $\widetilde{P_m}$ and P_e are in pu. Note that at the steady-state nominal condition when $\omega = \omega_0$, $\widetilde{P_m} = P_e$.

3.2.4 Small-signal swing equation

Considering small perturbations from an initial nominal condition notated by the subscript "$_0$", we have:

$$\omega = \omega_0 + \Delta\omega, \tag{3.13}$$

$$\widetilde{P_m} = \widetilde{P_{m0}} + \Delta P_m, \tag{3.14}$$

$$P_m = P_{m0} + \Delta P_m, \tag{3.15}$$

$$P_e = P_{e0} + \Delta P_e, \tag{3.16}$$

where $\widetilde{P_{m0}} = P_{m0} - k\omega_0^2 = P_{e0}$.

Then we have the following relationship according to (3.12).

$$LHS = 2H\frac{d\Delta\omega}{dt} \tag{3.17}$$

$$RHS = \underbrace{\widetilde{P_{m0} - P_{e0}}}_{=0} + \Delta P_m - \Delta P_e - D_1\Delta\omega = \Delta P_m - \Delta P_e - D_1\Delta\omega \tag{3.18}$$

The linearized swing equation is presented as follows.

$$2H\frac{d\Delta\omega}{dt} = \Delta P_m - \Delta P_e - D_1\Delta\omega. \tag{3.19}$$

In the Laplace domain, (3.19) becomes (3.21).

$$(2Hs + D_1)\Delta\omega = \Delta P_m - \Delta P_e. \tag{3.20}$$

With the swing equation given, we now investigate frequency responses for two scenarios. In the first scenario, a load is served by a single generator. In the second scenario, a generator is connected to a strong grid.

3.2.5 A stand-alone generator serving a load

For a stand-alone system with a generator serving a load with real power consumption notated as P_L, we may use (3.19) to investigate the frequency response of the system when it is subjected to a load increase.

Ignore all power losses in the electric system and assume that the mechanical power changes very slowly. For the time scale investigated, e.g., 10 seconds, the mechanical power does not vary, i.e., $\Delta P_m = 0$. Now consider a step response due to ΔP_L. First, we need to understand that $P_e = P_L$ for this system.

(3.19) becomes

$$2H\frac{d\Delta\omega}{dt} = -\Delta P_L - D_1\Delta\omega. \tag{3.21}$$

Steady-state response

The steady-state response of $\Delta\omega$ can be found by making the derivative of the speed deviation ($\Delta\dot{\omega}$) equal zero. (3.21) becomes:

$$-\Delta P_L - D_1\Delta\omega = 0. \tag{3.22}$$

If $\Delta P_L = 1$, then at steady-state, the speed deviation is $-\frac{1}{D_1}$.

The steady-state frequency deviation value can also be found from the transfer function. From (3.21), the transfer function from the load ΔP_L to the speed $\Delta\omega$ is as follows.

$$\frac{\Delta\omega}{\Delta P_L} = -\frac{1}{2Hs + D_1} \tag{3.23}$$

The time-domain steady-state response can be found by using the final value theorem:

$$\lim_{t\to\infty} f(t) = \lim_{s\to 0} sF(s) \tag{3.24}$$

where $f(t)$ is a time-domain function while $F(s)$ is its Laplace transform. We find:

$$\Delta\omega(t\to\infty) = \lim_{s\to 0} s\Delta\omega(s) = \lim_{s\to 0}\left(-\frac{1}{2Hs + D_1}\right) s\Delta P_L \tag{3.25}$$

If $\Delta P_L = 1/s$ for a step response, then we only need to examine the value of the transfer function in (3.23) at $s = 0$.

Based on this evaluation, we can also find that the steady-state frequency deviation is $-\frac{1}{D_1}$.

This indicates that an increase in load will cause a decrease in frequency. Moreover, D_1 is very small, which indicates a big decrease in frequency. Hence there is a need to develop control to reduce the steady-state frequency deviation. This is the task of primary frequency control or droop control.

Dynamic response

The dynamic response of $\Delta\omega(t)$ can be found through solving the first order differential equation in (3.21). In short, for $\Delta P_L = 1$,

$$\Delta\omega(t) = -\frac{1}{D_1}\left(1 - e^{-\frac{D_1}{2H}t}\right). \tag{3.26}$$

The dynamic response can also be evaluated from the Laplace transform.

$$\begin{aligned}\Delta\omega(s) &= -\frac{1}{2Hs + D_1}\Delta P_L(s) = -\frac{1}{2Hs + D_1}\frac{1}{s} \\ &= -\frac{1}{D_1}\left(\frac{1}{s} - \frac{1}{s + D_1/(2H)}\right)\end{aligned} \tag{3.27}$$

Inverse Laplace transformation indicates the time-domain expression of $\Delta\omega(t)$ as follows.

$$\Delta\omega(t) = -\frac{1}{D_1}\left(1 - e^{-\frac{D_1}{2H}t}\right) \tag{3.28}$$

3.2.6 Single-machine infinite-bus (SMIB) system

We now proceed to build a model for a SMIB system. A generator is connected to an infinite bus through a transmission line. The infinite bus refers to a large grid. This grid has infinitive inertia. The grid is represented by a voltage source with constant voltage magnitude and constant frequency. Change in load will not affect its frequency due to the infinitive inertia ($J \to \infty$ and $\dot{\omega} = 0$). The voltage phasor of the infinite bus is $V_\infty \angle 0$. The transmission line is represented by a pure reactance X_L.

Assuming the generator is represented by the simplest model as a voltage source ($E\angle\delta$) behind a reactance X_s and all electromagnetic dynamics are ignored, the electric power sending from the generator to the infinite bus is

$$P_e = \frac{EV_\infty}{X}\sin(\delta) \tag{3.29}$$

where $X = X_s + X_L$ is the total reactance, including the generator synchronizing reactance and line reactance.

E is the root mean square (RMS) value of the internal voltage proportional to the excitation current i_F on the rotor. δ is related to the rotor position θ in the following setup.

$$\begin{aligned}\theta &= \omega t + \theta_0 \\ &= \omega_0 t + \delta + \frac{\pi}{2}\end{aligned} \tag{3.30}$$

where θ_0 is the initial rotor position relative to a static reference.

The fact that the angle between the internal voltage's space vector and the synchronous reference frame is δ is explained in detail in Chapter 4. In this chapter, we will give a brief explanation.

δ is the position of the rotor's quadrature-axis (q-axis, at the position of $\theta - \frac{\pi}{2}$) relative to a rotating reference frame (at the position of $\omega_0 t$). This rotating reference frame has a constant nominal speed ω_0 and hence it is called a synchronous rotating reference frame. If the machine is rotating at the speed ω, and assuming that the direct axis (d-axis) is the rotor axis (the direction of the field generated by the excitation current i_F) while the q-axis is 90^0 lagging the d-axis in space, then the q-axis' position relative to the static reference is $\theta - \frac{\pi}{2} = \omega t + \theta_0 - \frac{\pi}{2}$.

As such we have

$$\delta = \theta - \frac{\pi}{2} - \omega_0 t = (\omega - \omega_0)t + \theta_0 - \frac{\pi}{2} \tag{3.31}$$

$$\dot{\delta} = \omega - \omega_0, \tag{3.32}$$

where δ is in radian (rad) while ω is in rad/s.

The dq-axes, θ, and δ are shown in Figure 3.1 for illustration.

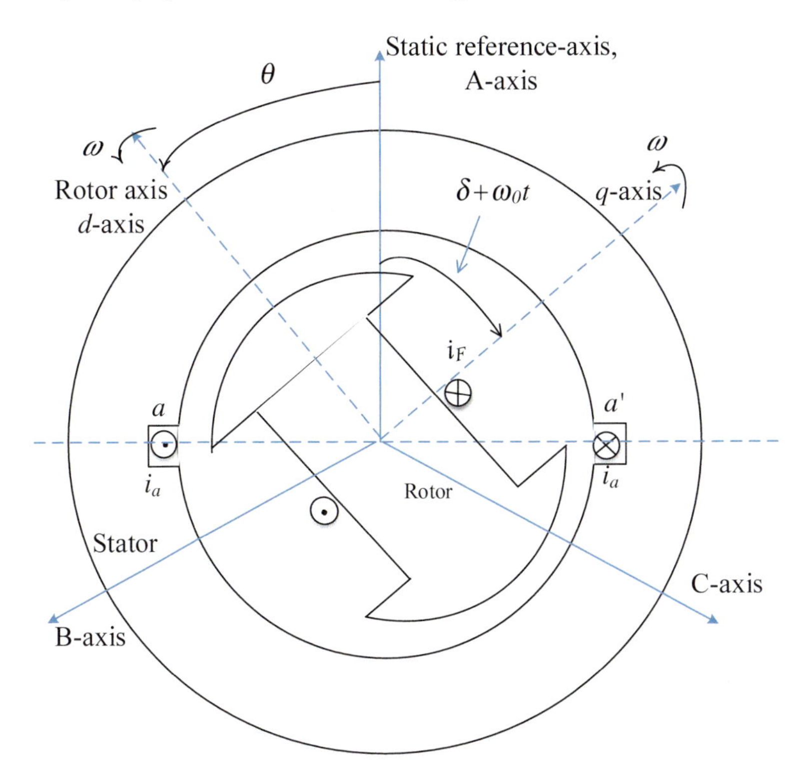

Figure 3.1: Illustration for dq-axes, θ and δ.

If we use per unit value for ω, we have

$$\dot{\delta} = \omega_0(\omega - 1). \tag{3.33}$$

where δ is in rad while ω is in per unit.

Note that for an angle, the physical unit is used. Nominal angle and normalized angle seem to have little meaning. Similarly, for time t, the unit is in seconds. Anderson and Fouad (2008) adopted a per unit value for t and that formulation becomes obsolete because it is very difficult to imagine a normalized time.

The small-signal model of (3.33) can be expressed in terms of $\Delta\delta$ and $\Delta\omega$:

$$\Delta\dot{\delta} = \omega_0 \Delta\omega. \tag{3.34}$$

The swing equations for a SMIB system are the following set.

$$\begin{aligned} \frac{d\delta}{dt} &= \omega_0(\omega - 1) \\ 2H\frac{d\omega}{dt} &= \widetilde{P_m} - P_e - D_1\Delta\omega, \end{aligned} \tag{3.35}$$

where $P_e = \frac{EV_\infty}{X}\sin(\delta)$.

Linearized model

The above swing equation set can be linearized at an equilibrium point or initial condition (ω_0, δ_0, P_{m0}, P_{e_0}).

$$\frac{d\Delta\delta}{dt} = \omega_0\Delta\omega \tag{3.36a}$$

$$2H\frac{d\Delta\omega}{dt} = \Delta P_m - \Delta P_e - D_1\Delta\omega. \tag{3.36b}$$

Applying a small perturbation for P_e using (3.6), we have

$$\Delta P_e = \underbrace{\frac{EV_\infty}{X}\cos(\delta_0)}_{T}\Delta\delta. \tag{3.37}$$

Replacing $\Delta\omega$ by $\Delta\dot{\delta}/\omega_0$ in (3.36b), we can obtain a second-order differential equation, with a single variable $\Delta\delta$.

$$\frac{2H}{\omega_0}\Delta\ddot{\delta} + \frac{D_1}{\omega_0}\Delta\dot{\delta} + T\Delta\delta = \Delta P_m. \tag{3.38}$$

In Bergen and Vittal (2009), two new parameters are defined:

$$M \triangleq \frac{2H}{\omega_0} \tag{3.39}$$

$$D \triangleq \frac{D_1}{\omega_0}. \tag{3.40}$$

The final second-order single-variable differential equation is

$$M\Delta\ddot{\delta} + D\Delta\dot{\delta} + T\Delta\delta = \Delta P_m. \tag{3.41}$$

In the Laplace domain, (3.41) becomes

$$(Ms^2 + Ds + T)\Delta\delta = \Delta P_m. \tag{3.42}$$

Steady-state frequency deviation and rotor angle deviation

Based on the swing equations (3.35), the steady-state value of the speed should be $\omega(t \to \infty) = 1$ since $\dot{\delta} = 0$ at steady-state. The steady-state angle should meet the requirement of $P_m - k\omega_0^2 = P_e = \frac{EV_\infty}{X}\sin\delta$. If the generator's prime mover increases its output P_m, then we should see an increase in the rotor angle δ.

If the increase is not significant, we can still use the linearized model to investigate $\Delta\delta$. According to $\Delta P_m = \Delta P_e = T\Delta\delta$ at steady-state, we should have $\Delta\delta = \frac{\Delta P_m}{T}$.

The transfer functions from ΔP_m to $\Delta\delta$ and $\Delta\omega$ are:

$$\frac{\Delta\delta}{\Delta P_m} = \frac{1}{Ms^2 + Ds + T} \tag{3.43}$$

$$\frac{\Delta\omega}{\Delta P_m} = \frac{s}{\omega_0(Ms^2 + Ds + T)} \tag{3.44}$$

If we substitute s by zero, we can find that the values of the two transfer functions are $1/T$ and 0. This is to say, if the mechanical power has a step response, the final angle will have $1/T$ increase in radus, while the frequency deviation will be zero, or the frequency will return to nominal after dynamics.

Remarks: The above investigation of a SMIB system shows that for a system with a strong grid, there are no frequency control issues. The SMIB frequency response case also confirms the second fact presented in the beginning of the chapter: at steady-state, frequency or speed everywhere is the same. Since the infinite bus keeps a nominal frequency, the generator's speed will be nominal at steady-state.

In cases related to real-world power system modeling, we should use infinite bus with discretion. To investigate the effect of frequency control, infinite bus should not be used to model a generator or a grid. This way of modeling enables realistic frequency response investigation.

In microgrids, power electronic converters are employed as the interfaces between distributed energy resources and the grid. Converters become the main control devices. Microgrids have two operating modes: grid-connected and autonomous. In the grid-connected mode, a microgrid is connected to a strong grid. While in the autonomous mode, a microgrid is a stand-alone system. In grid-connected mode, a converter does not need to provide a constant frequency since the grid will support the frequency. In the grid-connected mode, converters are usually set in PQ control mode. For example, a battery's charging or discharging power level will be set when it is plugged into a utility grid.

In the autonomous mode, e.g., a battery serving a load, the converter should consider frequency control. There has to be a way to regulate frequency like synchronous generators. Unlike synchronous generators, where frequency control is realized through turbine-governors, power electronic converters realize frequency control through converter control and modulation. The advantage is that converters can do control much faster while turbine-governors have slower responses. This could also be considered as a disadvantage that microgrids without conventional synchronous generators suffer significant frequency change due to lack of inertia.

Frequency and voltage control in microgrids will be addressed in Chapter 6.

3.3 How to reduce frequency deviation

From the previous analysis of a stand-alone system, we find that the steady-state frequency deviation is $-\Delta P_L/D_1$. D_1 is related to damping and is small, e.g., 1 pu. Hence, for 0.1 unit load change, the frequency change will be 0.1 pu or 6 Hz. In real-world, frequency deviation is to be kept within a tight range from the nominal 60 Hz. The lower frequency limit is set at 59.5 Hz according to North American Electric Reliability Corporation (NERC)'s operating guidelines North American Electric Reliability Corporation (2011). Small steady-state frequency deviation is achieved through primary frequency control.

3.3.1 Primary frequency control and its effect

Naturally, we can think of reducing frequency deviation by increasing D_1. However, D_1 is related to the friction of the mechanical system and it is also not energy efficient to increase friction. So we can use feedback control to achieve the similar effect. If the final closed-loop system in (3.23) becomes $\frac{\Delta\omega}{\Delta P_L} = -\frac{1}{2Hs+D_1+k}$, where k is a pure gain, then the steady-state frequency deviation becomes $1/(D_1+k)$. k can be set to achieve a small frequency deviation.

Simply, if we assume ΔP_m is the control point, by introducing a feedback from $\Delta\omega$ with a gain k, the closed-loop feedback system transfer function from ΔP_L to $\Delta\omega$ is

$$\frac{\Delta\omega}{\Delta P_L} = \frac{\text{Forward gain}}{1+\text{Loop gain}} = \frac{-1/(2Hs+D_1)}{1+k/(2Hs+D_1)} = \frac{-1}{2Hs+D_1+k} \tag{3.45}$$

where "Forward gain" means the transfer function block from ΔP_L to $\Delta\omega$ when the feedback loop is not considered.

We can see from the above equation that our mission of decreasing steady-state frequency deviation is accomplished.

In reality, we cannot directly treat ΔP_m as the control point. Instead, the mechanical power is produced after the turbine. The turbine is controlled by a governor mainly through a valve. If the valve is opened wider, more steam comes to blow the turbine and more mechanical power will be generated. The turbine and the governor can be modeled by a first-order system each respectively. The total transfer function is $\frac{1}{T_g s+1}\frac{1}{T_t s+1}$, where T_g is the time constant for a governor and T_t is the time constant of the turbine. The input to the block is a power reference P_c, the output is the mechanical power.

The power reference P_c is adjustable. Therefore P_c will be the control point. The frequency deviation is measured and amplified with a gain. This output will be used to modify the input into the governor as $P_c - k\Delta\omega$. Starting here, we will use the regulation parameter R as $1/k$. We will show that R works better when droop lines are plotted.

The block diagram consisting of the swing equation, turbine-governor block, and the droop control is presented in Figure 3.2.

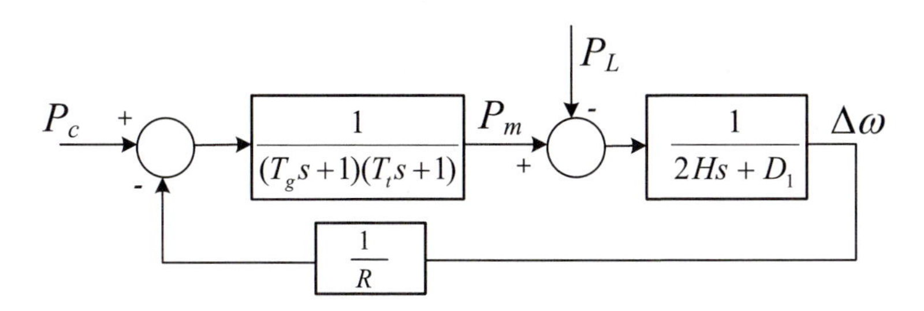

Figure 3.2: Block diagram of primary frequency (droop) control, turbine-governor and swing equation.

At steady-state, the relationship of P_m and $\Delta\omega$ is

$$P_m = P_c - \frac{1}{R}\Delta\omega \tag{3.46}$$

P_c will be changed by the secondary frequency control. If the secondary frequency control is not enabled, P_c is treated as a constant. Therefore, we have:

$$\Delta P_m = -\frac{1}{R}\Delta\omega \Longrightarrow \Delta\omega = -R\Delta P_m. \tag{3.47}$$

If we ignore the friction $D_1 = 0$, then the swing equation block becomes an integrator block. The input to the block should be 0, i.e., $\Delta P_L = \Delta P_m = -\frac{1}{R}\Delta\omega$. The final frequency deviation is

$$\Delta\omega = -R\Delta P_L. \tag{3.48}$$

This is the reason R instead of k is used, since we can quickly determine the physical meaning of R: with 1 pu load change, R unit frequency change will occur.

Regulation parameter R The unit of R is thus the unit of frequency divided by the unit of power. Therefore, R can be defined as 3 Hz per 100 MW or 5% pu with $S_b = 100$ MW. The meaning of the latter is given a 1 pu power change; the frequency change will be 5% pu or 5%pu $\times$ 60Hz/pu = 3 Hz.

Per unit is used for R most frequently. For generators, the power base is usually its own power rating. For example, a 1000 MW generator has a regulation parameter $R_1 = 5\%$ pu, and an 100 MW generator has a regulation parameter $R_2 = 5\%$ pu. They are referring to 1000 MW and 100 MW respectively. For the first generator, a change of 1000 MW will cause a 3 Hz frequency change. A change of 100 MW will cause a 0.3 Hz change. For the second one, the 100 MW change will cause a 3 Hz change. In the same power base of $S_b = 100MW$, then $R_1 = 0.5\%$ pu and $R_2 = 5\%$ pu.

Does mechanical power increase the same amount as the load increase? If we consider friction, these two are not exactly the same. Evaluating the control block diagram in Figure 3.2 at steady-state by replacing s with $j0$, i.e.:

$$\Delta P_m - \Delta P_L = D_1\Delta\omega = -D_1 R\Delta P_m \tag{3.49}$$

$$\Rightarrow (1 + D_1 R)\Delta P_m = \Delta P_L. \tag{3.50}$$

Let $D_1 = 1$, and $R = 0.05$, then $1.05\Delta P_m = \Delta P_L$.

Example Compute the approximate frequency drop after a generator of 1000 MW tripped in a big grid. This grid is assumed to have 1000 big generators each at 1000 MW and the droop regulation parameters are all 5% based on each generator's nominal power.

Solution: When one generator is tripped, the power imbalance is 1000 MW with more load than generation. Choosing the power base as 1000 MW, we should be able to compute the system's frequency deviation.

Ignoring the frictions of generators, then

$$\Delta P_L = \sum_{i=1}^{999} \Delta P_m = \sum_{i=1}^{999} \frac{-1}{R_i} \Delta\omega.$$

Therefore,

$$\Delta\omega = -\frac{\Delta P_L}{\sum_{i=1}^{999} 1/R_i} = -1/(999 \times 20) \text{ in pu} \tag{3.51}$$

$$= -60/(999 \times 20) \text{ in Hz} \approx -0.003\text{Hz} \tag{3.52}$$

In a real-world system such as the Eastern Interconnection, 59.997 Hz frequency indicates a significant power imbalance. If the system consists of 100 generators, the frequency deviation is approximately -0.03 Hz. This fact indicates that with more generators interconnected, frequency deviation can be reduced. This is one reason why the grid tends to be large-scale. In addition to the steady-state frequency deviation, dynamic frequency change can also be significantly reduced. For a system that has N generators, for a load change, each generator will share an average of $1/N$ power change. Examine the swing equation and replace $\dot{\omega}$ as $\frac{\Delta\omega}{\Delta t}$. We have:

$$2H\frac{\Delta\omega}{\Delta t} \approx 2H\frac{d\omega}{dt} = P_m - P_e = -\frac{\Delta P_L}{N}. \tag{3.53}$$

Here we assume Δt is very small and during that time scale, P_m is kept constant. The electric power from each generator shares the load increase. Therefore, $P_m - P_e = P_{m0} - (P_{e0} + \frac{\Delta P_L}{N}) = -\frac{\Delta P_L}{N}$.

Therefore, with more generators connected, the rate of frequency change can be reduced. This will be reflected in the reduction of maximum frequency deviation during the dynamics. For small systems, we tend to see big frequency excursion, while for big systems, frequency excursion during dynamics and steady-state will be insignificant. And this is the advantage of a big system.

3.3.2 Power sharing among multiple generators

Consider a system with n generators serving a load. First, let's set up the relationship between the frequency deviation versus the electric power P_e. At steady state, the following relationship should be true by making the speed dynamics $\frac{d\omega}{dt} = 0$.

$$0 = \Delta P_{mi} - \Delta P_{ei} - D_{1i}\Delta\omega_i, \quad ,i = 1, \cdots, n \tag{3.54}$$

This fact is very important: for a connected system, frequency or speed at steady-state should be the same everywhere. Therefore, the above relationship becomes:

$$0 = \Delta P_{mi} - \Delta P_{ei} - D_{1i}\Delta\omega. \tag{3.55}$$

by replacing ω_i with ω, the system frequency.

This tells us that the generator electric power increase is contributed by two elements: increase in the mechanical power ΔP_m, and the reduction of friction or the release of friction energy due to reduced speed $-D_1\Delta\omega$. Further, substituting ΔP_m using $-\frac{1}{R}\Delta\omega$, we have

$$0 = -\frac{1}{R_i}\Delta\omega - \Delta P_{ei} - D_{1i}\Delta\omega = -\underbrace{\left(\frac{1}{R_i} + D_{1i}\right)}_{\beta_i}\Delta\omega - \Delta P_{ei}. \tag{3.56}$$

Therefore, each generator's electric output change will be

$$\Delta P_{ei} = -\beta_i \Delta\omega. \tag{3.57}$$

Since all generators' output will contribute to the load, we have

$$\Delta P_L = \sum_i \Delta P_{ei} = -\sum_i \beta_i \Delta\omega \tag{3.58}$$

$$\Longrightarrow \Delta\omega = -\frac{\Delta P_L}{\sum_i \beta_i} \tag{3.59}$$

$$\Longrightarrow \Delta P_{ej} = \frac{\beta_j}{\sum_i \beta_i}\Delta P_L \tag{3.60}$$

$$\Longrightarrow \Delta P_{mj} = \frac{1/R_j}{\sum_i \beta_i}\Delta P_L \tag{3.61}$$

$$\Longrightarrow \Delta P_{e1} : \Delta P_{e2} : \cdots : \Delta P_{en} = \beta_1 : \beta_2 : \cdots : \beta_n \tag{3.62}$$

$$\Longrightarrow \Delta P_{m1} : \Delta P_{m2} : \cdots : \Delta P_{mn} = \frac{1}{R_1} : \frac{1}{R_2} : \cdots : \frac{1}{R_n}. \tag{3.63}$$

Figure 3.3 shows three generators with different droops. It can be seen that a shallow slope results in more power sharing.

3.3.3 Reactive power sharing

The concept of droop has been widely used for power sharing among generators and reactive power compensation devices. Droop control has also been

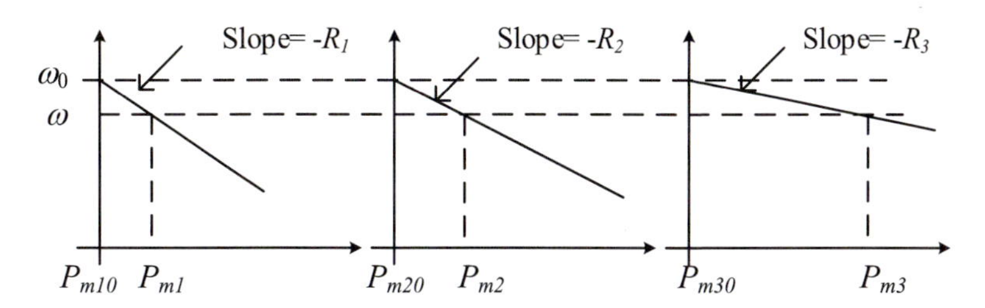

Figure 3.3: Effect of droop regulation parameters. $R_1 > R_2 > R_3$ and $\Delta P_{m1} < \Delta P_{m2} < \Delta P_{m3}$.

applied for power sharing among distributed energy resources in microgrids through power electronic converter control Chandorkar et al. (1993); Li and Kao (2009).

For reactive power sharing among Static Compensators (STATCOMs), V-I droop is adopted Hingorani and Gyugyi (2000). Each STATCOM is going to produce pure reactive power to the grid. And they are connected to the same system bus with voltage $\bar{V}_s$. The objective of Q sharing is to be able to let each STATCOM contribute a certain share. Since the system bus for each STATCOM is the same, the Q injection to the grid is completely dependent on the currents provided. When the system bus suffers a drop ΔV_s, it is expected that each STATCOM will increase its current and Q injection. Therefore, the relationship or droop control between ΔV_s and ΔI_i is as follows.

$$\Delta V_s = -K_i \Delta I_i \tag{3.64}$$

The contribution of $\Delta Q_i = \Delta(V_s \times I_i) \approx V_{s0}\Delta I_i$, assuming the voltage change is very small so a linear term $\Delta V_s I_{i0}$ is ignored.

$$\Delta Q_1 : \Delta Q_2 : \cdots : \Delta Q_n = \Delta I_1 : \Delta I_2 : \cdots : \Delta I_n = \frac{1}{K_1} : \frac{1}{K_2} : \cdots : \frac{1}{K_n}. \tag{3.65}$$

The droop control is realized through control blocks. Each STATCOM is in the voltage control mode to control the system voltage V_s. The order of the voltage control V_s^* should be generated through droop control:

$$V_s^* = V_{s0} - K_i I_i \tag{3.66}$$

where subscript $_0$ refers to the voltage setting.

3.4 How to eliminate frequency deviation

With droop control, the system will have a reduced steady-state frequency deviation. The next discussion is on how to eliminate the steady-state frequency deviation after a load change. The terminology in power systems is secondary frequency control and automatic generation control (AGC). In control, this is same as how to track a reference signal. The reference signal is the nominal frequency while the measurement is the system frequency. Tracking can be realized through feedback control.

3.4.1 How to track a signal

DC signal tracking

To track a DC signal, we can simply use an integrator, as shown in Figure 3.4. The error $e = r - y$ is sent to an integrator to generate the control action u. Suppose the plant model is simplified as a gain $1/\beta$. The output y is related to the error and the reference as

$$\frac{y}{e} = \frac{y}{r-y} = \frac{k}{\beta s} \tag{3.67}$$

$$\frac{y}{r} = \frac{k/(\beta s)}{1+k/(\beta s)} = \frac{1}{1+\frac{\beta s}{k}} \tag{3.68}$$

where k is the gain of the integral control. The steady-state gain from r to y is 1, i.e., at steady-state, $y = r$.

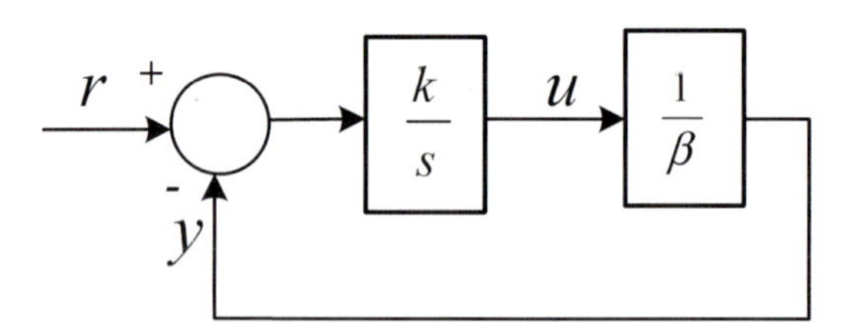

Figure 3.4: Use integral control to track DC signals.

Integral control is a widely used in control applications to track DC signals. In AGC, integral control is used. In power electronic converters, integral control is used to track current references, voltage references and power references.

Dynamic perspective From the closed-loop transfer function from r to y (3.68), it can be observed that the time constant of the system is β/k. Therefore, a large k indicates a faster response. k, as a gain, has a limit. In Figure 3.4, the plant model is a simplified model with a pure gain. In reality, the plant model also has dynamics. When k is too large, interactions with the plant dynamics may cause undesired performance, even instability. Below is an example where the plant is a first order system $\frac{1}{1+\tau s}$. The closed-loop system transfer function is

$$\frac{y}{r} = \frac{\frac{k}{s}\frac{1}{1+\tau s}}{1+\frac{k}{s}\frac{1}{1+\tau s}} = \frac{1}{\frac{s(1+\tau s)}{k}+1} = \frac{1}{1+s/k+\tau s^2/k} \tag{3.69}$$

When k assumes a large number, the system will have poor damping. Figure 3.5 is the step response of a system with $\tau = 1$ and $k = 100$. The poles of the system are $-0.5 \pm j10$. The damping ratio is defined as $\frac{\sigma}{\omega}$ where σ and ω are the real part and imaginary part of a pair of complex conjugate poles or eigenvalues $-\sigma \pm j\omega$.

For the above system, the damping ratio is 5%. This is due to a large gain of the integrator since the polynomial in the denominator can be written as $\frac{\tau}{k}\left(s^2 + s/\tau + k/\tau\right)$. The poles of the system or the roots for this polynomial are

$$s = -\frac{1}{2\tau} \pm j\sqrt{\frac{k}{\tau}} = \sigma \pm j\omega.$$

The poles decide the dynamic response of the closed-loop system. A large gain k results in a large resonance frequency ω and hence a small damping ratio. In general, a closed-loop system transfer function from the error to the reference signal can be expressed as

$$\frac{e}{r} = \frac{1}{1+L(s)} \tag{3.70}$$

where $L(s)$ is the loop gain, which is the transfer function from the beginning of the loop to the end of the loop, with the loop disconnected.

To make y track r when r is a DC signal is to make e go to zero for DC input. This translates to make e go to zero when $s = j0$.

Examine when $L(s)$ has an integrator and can be expressed as

$$L(s) = \frac{1}{s}G(s) \tag{3.71}$$

where $G(s)$ has a nonzero steady-state gain.

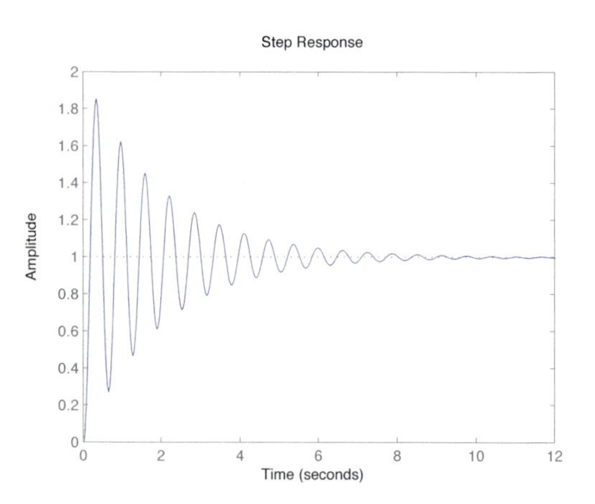

Figure 3.5: Step response for $1/(1 + 0.01s + 0.01s^2)$.

For this scenario, we can be sure that the loop gain has an infinite gain when $s = j0$. In turn, the transfer function from r to e will have a zero gain for DC signals (or when $s = j0$). That is, y can track r when r is a DC signal.

When the reference signal is not a DC signal, integral control may not work. For example for a ramp signal t with a Laplacian expression as $1/s^2$, a double integrator should be used to track the signal Aström and Murray (2010).

Sinusoidal signal tracking

Most of the time, in power systems, we would like to track ac signals. For example, how to maintain a 60 Hz current waveform with the desired magnitude, frequency and phase angle? An open-loop control, e.g., the sinusoidal pulse width modulation (PWM) switching, can realize the objective. Here the discussion is limited to feedback control.

Making three-phase currents track reference ac signals has been seen in power electronic converter control Yazdani and Iravani (2010). There are two approaches to get a desired current waveform through feedback control. The first is to convert the *abc* signals to signals viewed from a rotating reference frame called *dq*-reference frame. In *dq*, the sinusoidal currents are all constants at steady-state. With DC signals, tracking is easy. We can again rely on integral control or proportional integral (PI) control. Examples on *dq*-based converter control can be found in Chapter 6.

The second approach is to directly track an ac signal at 60 Hz. This type of control is called resonant control. With proportional block added, the control has a name of proportional resonant (PR) control. We will explain resonant control using the analogy of DC signal tracking.

To track a DC signal, we would like the loop gain, go to infinity when $s = j0$. Then the error between the reference and the output will be zero. To track an ac signal of certain frequency ω, then the loop gain should go to infinity when $s = j\omega$. Therefore, a resonant controller $\frac{1}{s-j\omega}$ will fulfill the task. $\frac{1}{s-j\omega}$ is not a realistic controller. $\frac{s}{s^2+\omega^2}$ will also make the gain at $s = j\omega$ goes to infinity. A practical resonant controller will have some damping and the transfer function will be

$$C(s) = \frac{ks}{s^2 + 2\alpha\omega s + \omega^2} \tag{3.72}$$

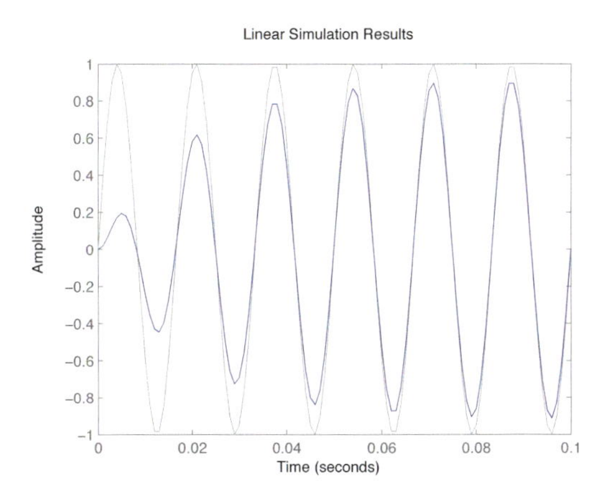

Figure 3.6: Tracking a 60 Hz signal. Plant model $P(s) = 0.5$, controller $C(s) = 100s/(s^2 + 7.54s + 377^2)$.

Figure 3.6 shows the reference signal and the output signal. The plant model is represented by a constant 0.5. A resonant controller is employed. The linear system is simulated using MATLAB command *lsim*. Figure 3.6 was generated by the following code.

```
s = tf('s');
k = 200;
a = 0.01;
w = 2*pi*60;
P = 0.5;
C = k*s/(s^2+ 2*a*w*s +w^2);
```

```
G = feedback(P*C,1);
t =0:0.001:0.1;
u = sin(377*t);
lsim(G,u,t);
```

3.4.2 Secondary frequency control

Secondary frequency control applies every two minutes. The system first sees frequency deviation, then secondary frequency control brings the system frequency back to nominal. As a control problem, this problem is to make the frequency track a nominal frequency ω^*. Naturally, integral control can be applied. The controller output should be P_c, the power setting of the governor. The block diagram of a generator serving a load is presented in Figure 3.7. Figure 3.7 is a control block diagram with a more realistic flavor.

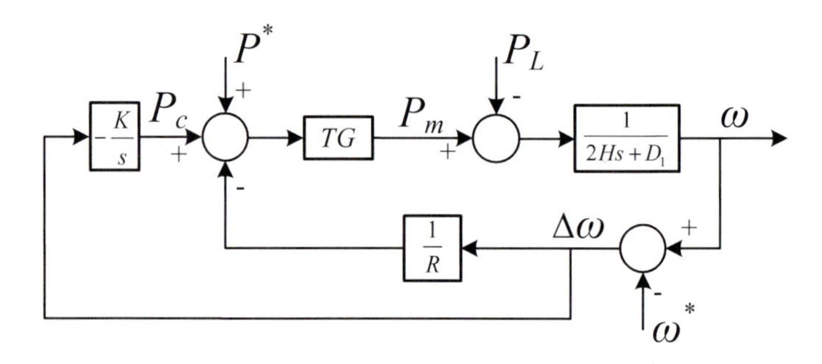

Figure 3.7: Block diagram of secondary frequency control.

For linear system analysis, we use a small-signal model as Figure 3.8.

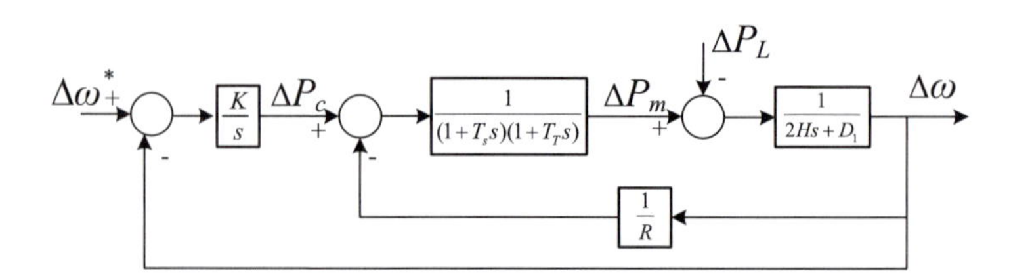

Figure 3.8: Block diagram of secondary frequency control for small-signal analysis.

Steady-state analysis

From the control block diagram in Figure 3.7, it can be seen that at steady-state, $P_m = P_c - \frac{1}{R}\Delta\omega$. In addition, with secondary frequency control, the input to the integrator will be 0 at steady-state. Therefore, at steady-state, $P_m = P_c$.

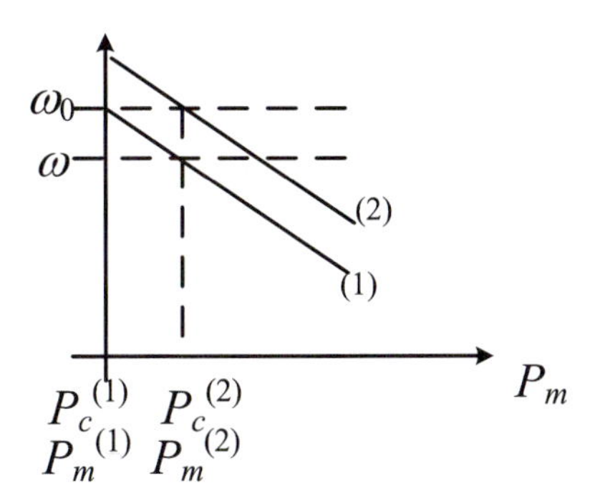

Figure 3.9: Effect of secondary frequency control.

The effect of the secondary frequency control can be explained using Figure 3.9. Initially the system is operating at the nominal condition and the setting of the reference power is $P_c^{(1)}$. When the system has a load increase, with only primary frequency control, the system will see a frequency drop. The frequency is now ω at steady-state. The corresponding mechanical power output is $P_m^{(2)}$. P_c will not change. With secondary frequency control enacted, the system's frequency will be brought back to nominal. However, the mechanical power should be kept at $P_m^{(2)}$. This is achieved by moving the droop line upward to (2). Then at $P_m^{(2)}$, the system frequency becomes nominal. At nominal frequency, the reference power and the mechanical power are the same. Therefore, in Figure 3.9, $P_m^{(2)}$ is shown as $P_c^{(2)}$. The effect of the secondary frequency control is to increase the reference power from $P_c^{(1)}$ to $P_c^{(2)}$.

The steady-state gain to be examined is the one from $\Delta\omega^*$ to $\Delta\omega$. A simplified way to examine the gain is to first obtain the steady state gain from ΔP_c to $\Delta\omega$. The steady-state gain is $\frac{1}{D_1+1/R}$ or approximately R as shown in Figure 3.10.

Second, examine the closed-loop system in Figure 3.10. The transfer function is now:

$$\frac{\Delta\omega}{\Delta\omega^*} = \frac{RK/s}{1+RK/s} = \frac{1}{1+s/(RK)} \tag{3.73}$$

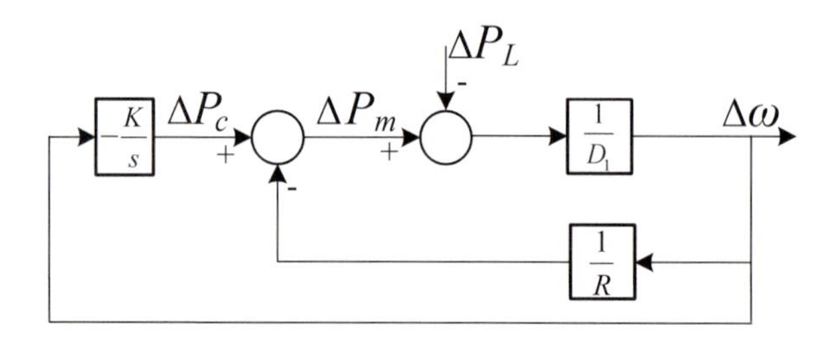

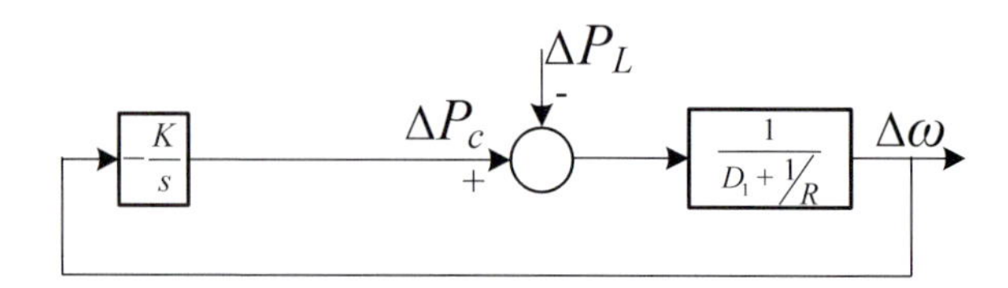

Figure 3.10: Simplified block diagram of secondary frequency control for steady-state computation. Except for the integrator, all other $s = j0$.

Obviously, the steady-state gain is 1. Therefore, ω can track ω^*.
The transfer function from ΔP_L to $\Delta\omega$ is as follows.

$$\frac{\Delta\omega}{\Delta P_L} = \frac{-R}{1 + RK/s} \tag{3.74}$$

When $s = j0$, the gain is zero. This means that even if there is a load change, the frequency will stay at nominal. The above analysis is based on simplified steady-state gain. One can find the complete closed-loop transfer function and substitute $s = j0$ to find the same result.

How to select the integral control gain k

Based on the close-loop system of the simplified model (3.73), to ensure stability of the closed-loop system, the poles of the system should be located in the left half plane. The pole is $-RK$. Therefore $K > 0$.

A major principle of control design is for different functions/loops to have different bandwidths. Bandwidth is defined as the frequency at which a transfer function's magnitude becomes -3 dB if its steady-state gain is 1. Consider a first-order system $G(s) = \frac{1}{\tau s+1}$. Then

$$|G(j\omega)| = \left|\frac{1}{j\omega\tau + 1}\right|.$$

When $\omega = 1/\tau$, $|G(j\omega)| = \frac{1}{\sqrt{2}} = 20\log_{10}\frac{1}{\sqrt{2}}$ dB $= -3$ dB.

In this case, the closed-loop system with integral control approximately has a bandwidth of RK. This bandwidth should be much lower than the bandwidth of the plant that the design is based on. The plant includes the droop control. If we assume the turbine-governor block is 1, the close-loop system from ΔP_c to $\Delta\omega$ has a transfer function of $\frac{1}{2Hs+1/R}$ assuming $1/R \gg D_1$. Therefore, the bandwidth is approximately $1/(2HR)$.

Use $H = 5$, $R = 0.05$, the two bandwidths are 2 rad/s for the inner loop, and $0.05K$ for the outer loop. To be much lower than 2 rad/s, $K = 1$ results in a reasonable bandwidth of 0.05 rad/s. We will expect the time for the integral control to bring the frequency back to 60 Hz is approximately in the order of several units of 20 seconds.

Stability analysis

In this section of Frequency Control, we will discuss a little bit of stability analysis. Stability issues can be met during control design. Very frequently, a large gain can make a system unstable. As an example, let us examine the single generator serving a load case where the integral control gain should have a limit. In Figure 3.16, the gain is chosen to be 1. If the gain is 3, the system becomes unstable, see Figure 3.11.

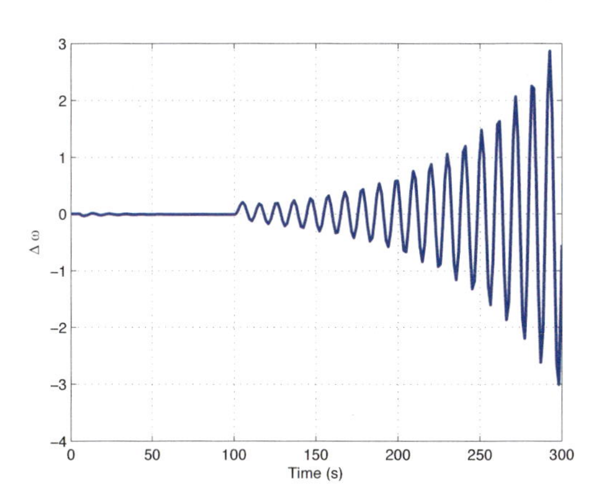

Figure 3.11: A system becomes unstable due to a large integral control gain.

System stability or instability can be identified using many stability criteria, e.g., Routh–Hurwitz (if the system's characteristic function and the polynomial's coefficients are given), system eigenvalues (if the system state-space model system matrix is given or transfer function is given), root loci (if the open-loop transfer function is given), Nyquist stability criterion and

Bode plots developed from Nyquist criterion (if the open loop transfer function is given).

For engineering analysis, root loci and Bode plots are frequently used. Though both can be implemented when the loop gain is known, they do have some subtle differences. Using root loci, we can exactly tell the limit of the gain. While using Bode plots, we have to come up with a loop gain corresponding to certain gain, and check the gain margin at −180 degrees. Bergen and Vittal (2009) adopts root loci frequently to check stability.

For the system in Figure 3.8, opening the secondary frequency control loop, the loop gain transfer function is as follows.

$$\begin{aligned} L_1(s) &= \frac{K}{s}\frac{\frac{1}{T_T s+1}\frac{1}{T_s s+1}\frac{1}{2Hs+D_1}}{1+\frac{1}{R}\frac{1}{2Hs+D_1}\frac{1}{T_T s+1}\frac{1}{T_s s+1}} = \frac{K}{s\left((T_T s+1)(T_s s+1)(2Hs+D_1)+1/R\right)} \\ &= KL(s) \end{aligned} \tag{3.75}$$

The root loci plot is generated by calling a MATLAB function *rlocus* and is shown in Figure 3.12. Root loci show the closed-loop system poles with a varying gain. When the gain is $K \to 0$, the closed-loop system poles and the open-loop system poles are the same. When the gain is $K \to \infty$, the closed-loop system's poles will approach to the open-loop system's zeros or infinity. This can be explained by examining the following two closed-loop transfer functions:

$$G_1 = \frac{L}{1+KL} \tag{3.76a}$$

$$G_2 = \frac{1}{1+KL}. \tag{3.76b}$$

When $k \to 0$, $G_1 \approx L$. Hence, G_1's poles are L's poles.

When $k \to \infty$, $G_2 = \frac{\frac{1}{KL}}{\frac{1}{KL}+1} \approx \frac{1}{KL}$. Hence, G_2's poles are L's zeros.

It can be seen that the limit for k is 2.7. That is why when the gain is chosen to be 3, this system ends up with instability.

For the root locus method, readers can refer classic control texts, e.g., Dorf and Bishop (1998). For the above loop gain $L(s)$, there are 4 poles (a pair of complex conjugate poles and two poles on the real axis). There is no zero.

The four poles are: 0, −2.192, $-0.054 \pm j0.6167$.

For negative feedback systems, root loci are located on the real axis as long as the total number of real poles on the right of the root loci position

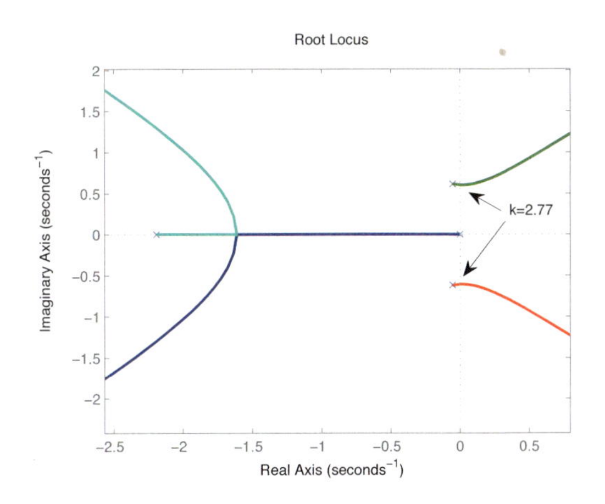

Figure 3.12: Root loci plot for $\frac{1}{s((T_T s+1)(T_s s+1)(2Hs+D_1)+1/R)}$.

is odd. In this case, the two poles on the real axis are 0 and -2.192. The real axis between 0 and -2.192 are root loci.

There should be $n-m$ asymptotes, where n is the number of poles and m is the number of zeros. The angle of the asymptotes with respect to the real-axis is $(2k+1)180°/(n-m)$, where k is an integer number. In this case, there will be four asymptotes and the angles are $\pm 45°$ and $\pm 135°$.

The asymptote centroid or the center of the four asymptotes can be computed using poles and zeros:

$$s_0 = \frac{\sum p_i - \sum z_i}{n-m} \quad \text{center-of-gravity rule,} \tag{3.77}$$

where p_i are the poles and z_i are the zeros, n is the number of the poles and m is the number of zeros.

In this case, $s_0 = -0.575$.

```
Tt = 5; Ts = 0.5; H = 5; D1 = 1; R = 0.05;
s = tf('s');
L = 1/s*1/(1/R +(Tt*s+1)*(Ts*s+1)*(2*H*s+D1));
rlocus(L);
```

Note that here we assume a negative feedback system and L is the loop gain of the negative feedback system when we open the loop. If the closed-loop system is a positive feedback loop system, then we should adopt $rlocus(-L)$ to plot the root loci for a positive feedback system.

3.4.3 Bring tie-line power flow schedule back to the original

When there are many areas connected together, tie-line power flow schedules are preferred to be constant as scheduled. Keeping tie-line power flow schedule constant is the same as requiring each area to take care of its own load change. Naturally, this task can also be realized through integral control.

Area control error (ACE)

The combined signal to the integral control used for AGC is $k\Delta\omega + \Delta P_{tie}$. Any k will take the frequency back to nominal and the tie-line schedule back to the schedule one. The next question is: Can we make this signal reflect the power imbalance in each area so we can immediately know which area is short of generation and which area has no change in power? All computation is based on steady-state variables after droop control but before AGC is applied.

How do we make the control signal, a combination of frequency deviation and tie-line deviation reflect power imbalance?

The derivation below shows that the ACE signal can be expressed by ΔP_m and ΔP_L of an area.

$$\begin{aligned} k\Delta\omega + \Delta P_{tie} &= -kR\Delta P_m + \Delta P_e - \Delta P_L \\ &= -kR\Delta P_m + \beta R\Delta P_m - \Delta P_L \\ &= (\beta - k)R\Delta P_m - \Delta P_L. \end{aligned} \tag{3.78}$$

Here we used the relationship between ΔP_e and ΔP_m: $\Delta P_e = \beta R\Delta P_m$. Therefore, when $k = \beta$, the signal becomes $-\Delta P_L$. This signal is defined as Area Control Error (ACE).

$$ACE = \beta\Delta\omega + \Delta P_{tie} \tag{3.79}$$

When the system is only equipped with primary frequency control and the system is at the steady-state,

$$ACE = -\Delta P_L. \tag{3.80}$$

ACE reflects the power imbalance in each area. Figure 3.13 shows the control block diagram with ACE control.

Example of ACE

For a three-area system, if Area 1 has a load increase, the entire system will experience a frequency increase due to the primary frequency control

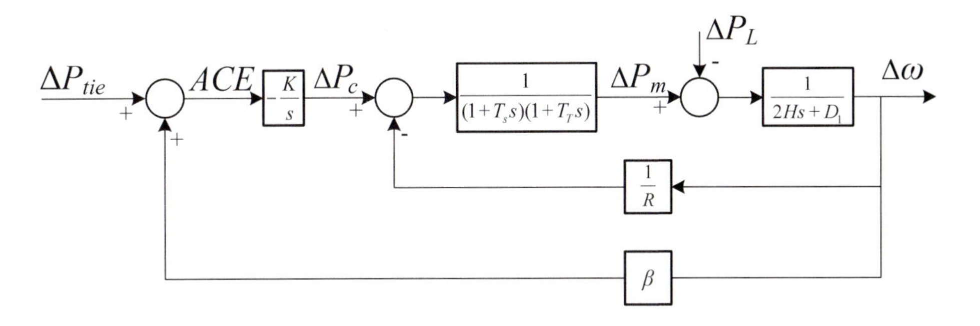

Figure 3.13: Automatic generator control and ACE.

installed on the generators. Suppose that each generator is 100 MW with 5% droop. The load increase in Area 1 is 100 MW. Compute ACEs for three Areas.

Answer: The the frequency drop is 1 Hz since

$$\Delta\omega = \frac{-\Delta P_L}{\sum \frac{1}{R_i}} = -\frac{1}{\frac{1}{0.05} + \frac{1}{0.05} + \frac{1}{0.05}}\text{pu} = \frac{-1}{60}\text{pu} = -1\text{Hz}.$$

Area 1 mechanical power will have $\frac{-1}{0.05} \times \frac{-1}{60} = \frac{1}{3}$ pu increase. Assume the friction is ignored and $P_m = P_e$. Since there is a load increase of 1 pu, the tie-line flow will decrease by $\frac{2}{3}$ pu ($\Delta P_{tie,1} = \Delta P_{e1} - \Delta P_{L1}$). The ACE signal is $\beta_1 \Delta\omega + \Delta P_{tie,1} = 20 \times \frac{-1}{60} - \frac{2}{3} = -1$ pu.

On the other hand, for Area 2, the mechanical power will have $\frac{1}{3}$ pu increase. Since there is no load change, this change is shifted to the tie-line flow change completely. Therefore, $\Delta P_{tie,2} = \frac{1}{3}$. $ACE_2 = 20 \times \frac{-1}{60} + \frac{1}{3} = 0$. Similarly, we can find $ACE_3 = 0$.

This example shows that an ACE signal reflects the power imbalance.

3.5 Validation of Frequency Control Design

3.5.1 A single generator serving a load

To validate the control effect in Figure 3.8, we can build models in MATLAB/Simulink for a system with a single generator serving a load, shown in Figure 3.14. To validate the control effect of ACE in Figure 3.13, a more sophisticated block diagram representing multi-areas and tie-line flow should be built. The MATLAB/Simulink block diagram of Figure 3.8 is shown in Figure 3.15.

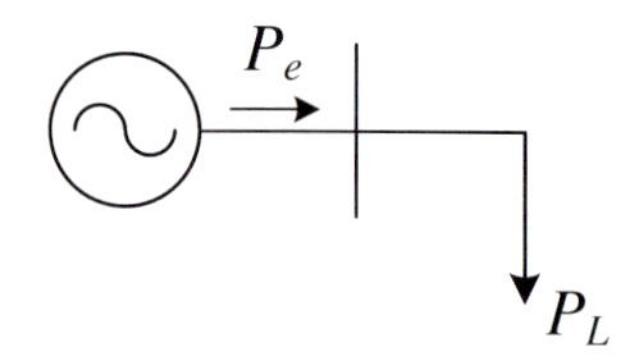

Figure 3.14: A single generator serving a load.

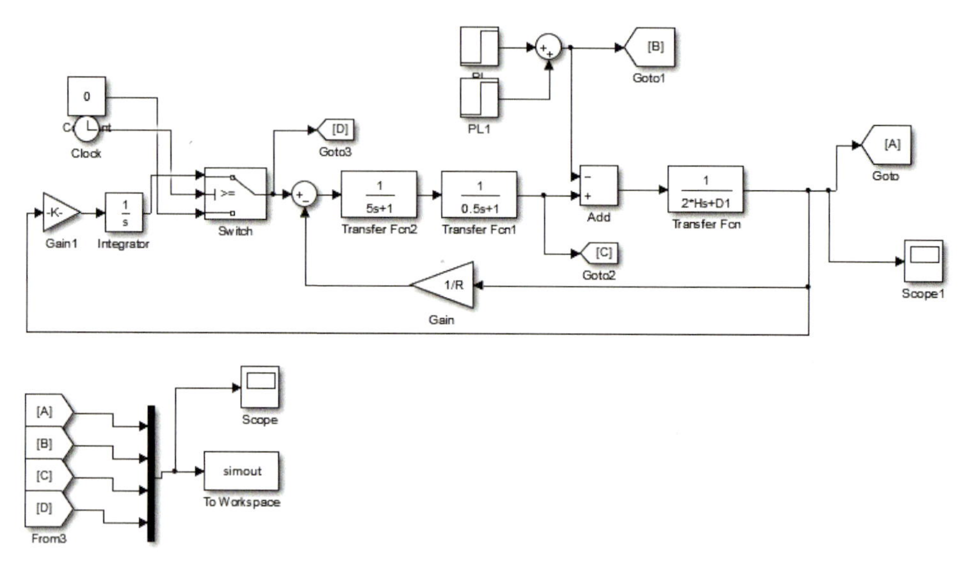

Figure 3.15: Simulink screen shot. $H = 5, R = 0.05, D_1 = 1, k = 0.5$.

The Simulink model has two step responses from the load. At 10 seconds, the load has a 0.2 pu increase. At 200 seconds, another 0.2 pu increase happened. A switch is employed to enable the secondary frequency control only after 100 seconds.

The simulation results in Figure 3.16 show that from 10–100 seconds, droop control works and the frequency achieves steady state. The frequency deviation is $\Delta\omega = -\Delta P_L/(1/R + D_1) = -0.2/21$. The mechanical power ΔP_m in turn will have an increase of $\frac{1/R}{1/R+D_1}\Delta P_L = \frac{20}{21} \times 0.2$, slightly less than ΔP_L.

At 100 seconds, the secondary frequency control starts to work and brings the frequency back to nominal. The frequency deviation will become 0 at steady-state. Note that the steady-state mechanical power ΔP_m will be the same as the load: $\Delta P_m = \Delta P_L = 0.2$ pu. The power order ΔP_c will be changed to match the load: $\Delta P_c = \Delta P_L = 0.2$ pu. Note that the steady-state has not been reached at 200 seconds in Figure 3.16.

At 200 seconds, there is another load increase. Both droop control and secondary frequency control work together and the frequency achieves nominal after dynamic transients. Both ΔP_c and ΔP_m will match ΔP_L and are now 0.4 pu: $\Delta P_c = \Delta P_m = \Delta P_L = 0.4$ pu.

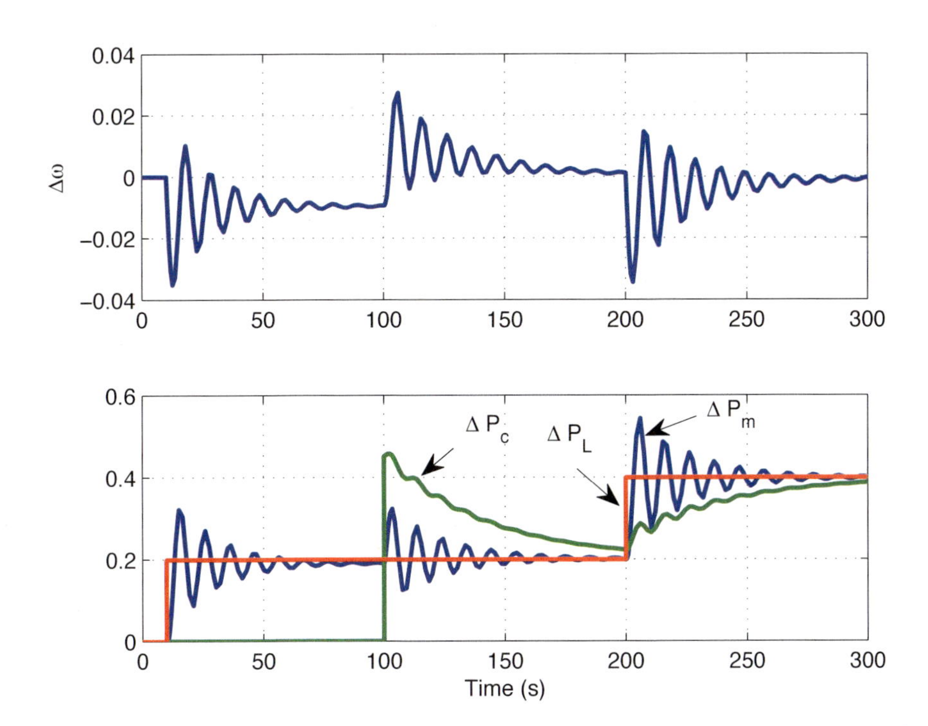

Figure 3.16: Simulation plots. At $t = 10$ s, the load has a 0.2 pu increase. At $t = 100$s, secondary frequency control is enabled. At $t = 200$ s, another 0.2pu increase in load occurs.

3.5.2 Two generators serving a load

Using the example system shown in Figure 3.17, we try to show the dynamic response of the system due to an increase in load and how generators share the load increase with droop control and further with secondary frequency control.

Steady-state power sharing among generators can be found based on

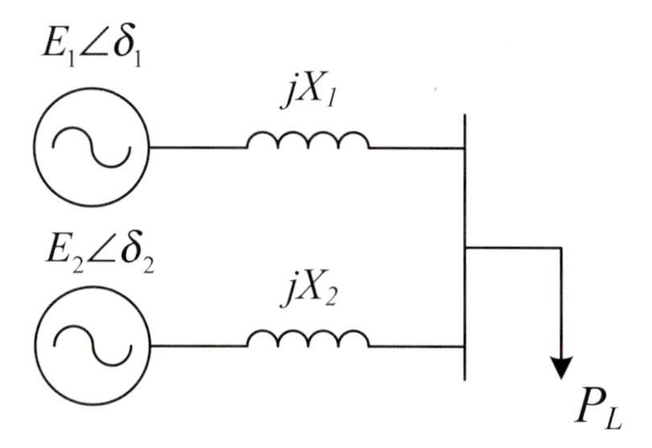

Figure 3.17: Two generators serving a load.

droop parameters:

$$\Delta P_{e,i} = \frac{\beta_i}{\sum_i \beta_i}\Delta P_L. \tag{3.81}$$

Therefore, if the two generators are having the same droop parameters, they will each share 50% of the load increase. However, what is the initial response of each generator? This is an interesting question. At the moment when a load has a sudden change, how does each generator share the load change? This is determined by circuit characteristics. Dynamic model building is important to let us understand the process.

It is not vital to express each generator's P_{ei} by P_L and the rotor angles δ. With this expression, we can have small-signal model of ΔP_e related to ΔP_L and $\Delta\delta$.

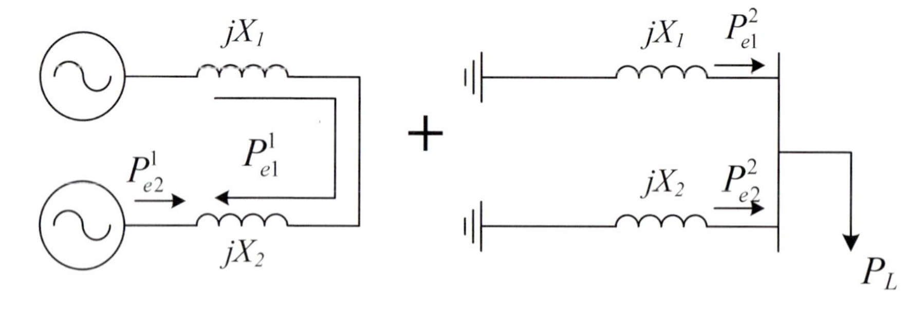

Figure 3.18: The system with two generators serving a load is decomposed into two systems.

Superposition is a straightforward way to obtain the expression of P_{ei}. We will have two circuits, one with the two voltage sources but without the load while the other is with the load but the voltage sources are zero as shown in Figure 3.18.

Note that load can be treated as a current injection or voltage source. So the above superposition is accurate. Therefore, the electric power from each generator has two components.

$$P_{e1} = P_{e1}^1 + P_{e1}^2 = \frac{E_1E_2}{X_1+X_2}\sin(\delta_1-\delta_2) + \frac{X_2}{X_1+X_2}P_L \tag{3.82}$$

$$P_{e2} = P_{e2}^1 + P_{e2}^2 = \frac{E_1E_2}{X_1+X_2}\sin(\delta_2-\delta_1) + \frac{X_1}{X_1+X_2}P_L \tag{3.83}$$

The linearized expressions are as follows.

$$\Delta P_{e1} = \frac{E_1E_2}{X_1+X_2}\cos(\delta_1^0-\delta_2^0)(\Delta\delta_1-\Delta\delta_2) + \frac{X_2}{X_1+X_2}\Delta P_L \tag{3.84}$$

$$\Delta P_{e2} = \frac{E_1E_2}{X_1+X_2}\cos(\delta_2^0-\delta_1^0)(\Delta\delta_1-\Delta\delta_2) + \frac{X_1}{X_1+X_2}\Delta P_L \tag{3.85}$$

If we define a T and let $T \triangleq \frac{E_1E_2}{X_1+X_2}\cos(\delta_1^0-\delta_2^0)$. Then

$$\Delta P_{e1} = T(\Delta\delta_1-\Delta\delta_2) + \frac{X_2}{X_1+X_2}\Delta P_L \tag{3.86}$$

$$\Delta P_{e2} = -T(\Delta\delta_1-\Delta\delta_2) + \frac{X_1}{X_1+X_2}\Delta P_L \tag{3.87}$$

The above analysis answers the question regarding initial response of each generator's power sharing. Since rotor angles cannot change from t_0^- to t_0^+, then at the moment of load increase, each generator will share the load increase inversely proportional to its reactance, or $\frac{\Delta P_{e1}}{\Delta P_{e2}} = \frac{X_2}{X_1}$.

Simulink blocks The screen shot simulink blocks are shown in Figure 3.19. Note that in this case, we will add an integrator for each generator to obtain $\Delta\delta$ (in radian). Three case studies are conducted.

1. Case 1: $R_1 = 0.05, R_2 = 0.1, K_1 = K_2 = 1$. The simulation results are shown in Figure 3.20. This case shows that Gen 1 shares more power due to a less R for primary frequency response. However, after secondary frequency control, the power sharing is the same since the gains of the integral control are the same.

2. Case 2: $R_1 = 0.05, R_2 = 0.1, K_1 = 1, K_2 = 0.5$. The simulation results are shown in Figure 3.21. In Case 2, Gen 1 (Green line) shares more after the secondary frequency control since a greater integral control gain is used.

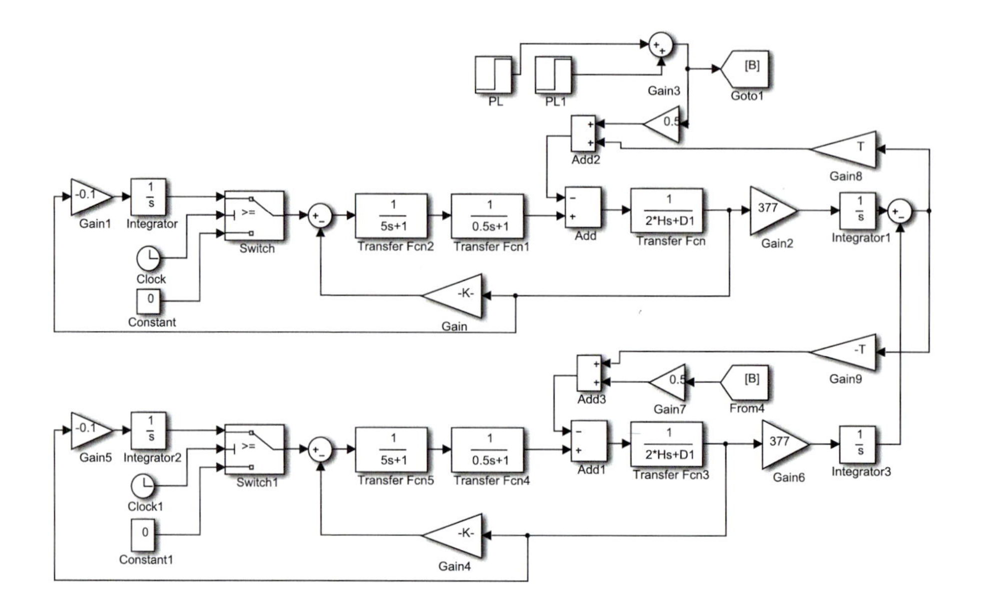

Figure 3.19: Simulink blocks for the system of two generators serving a load. $H = 5, D_1 = 1, T = 1$.

3. Case 3: $R_1 = 0.05, R_2 = 0.1, K_1 = 0.5, K_2 = 1$. The simulation results are shown in Figure 3.22. In Case 3, Gen 2 (red line) shares more after the secondary frequency control since a greater integral control gain is used.

The dynamic simulation results give a detail insightful look at the system behavior. There are a couple of things to be noticed.

- The rotor angles can keep reducing. This is due to the fact that without a secondary frequency control, frequency can be kept below 60 Hz for a period of time. During that period, the angles keep reducing. Therefore, behavior of angles is usually not used for stability judgement. In this case study, the system is obviously stable. The reason is that the linearized system with two rotor angles and two speeds as state variables always has an eigenvalue as 0. Instead, the rotor angle difference will be used to examine stability. Figure 3.23 shows the dynamic behavior of the rotor angle difference and each generator's electric power output.

  Note that at the moment of the load change, each generator shares the

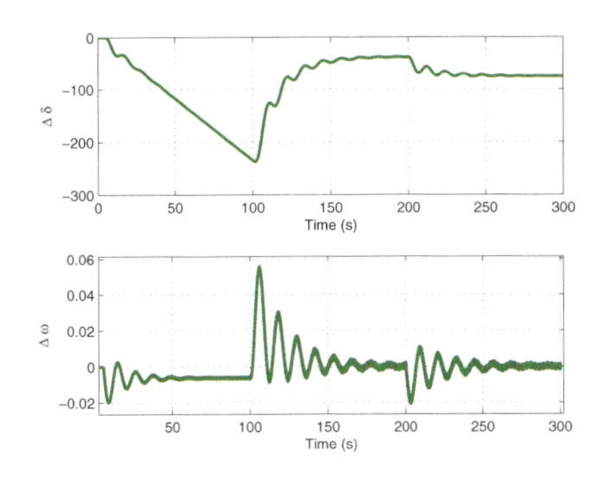

(a) Case 1: Rotor angle and speed. $R_1 = 0.05, R_2 = 0.1, k_1 = k_2 = 1.$

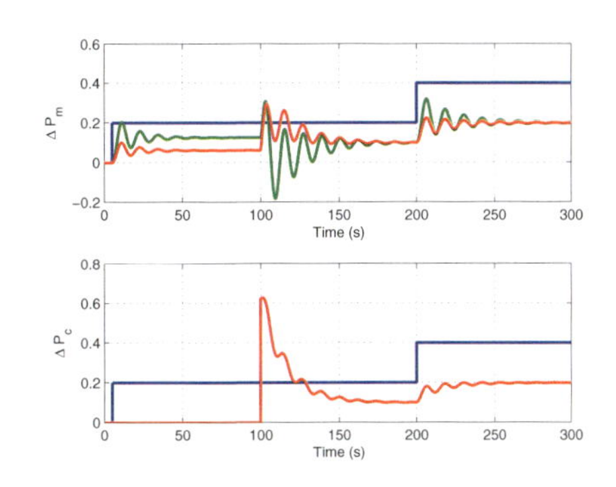

(b) Case 1: ΔP_m and ΔP_c. $R_1 = 0.05, R_2 = 0.1, k_1 = k_2 = 1.$

Figure 3.20: Case 1 simulation results for the system of two generators serving a load.

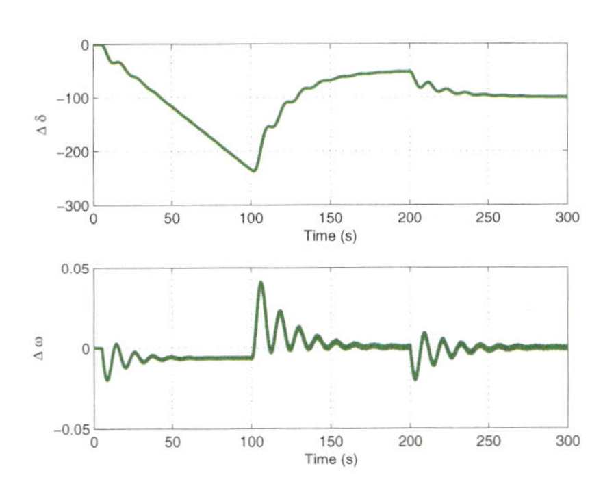

(a) Case 2: Rotor angle and speed. $R_1 = 0.05, R_2 = 0.1, k_1 = 1, k_2 = 0.5.$

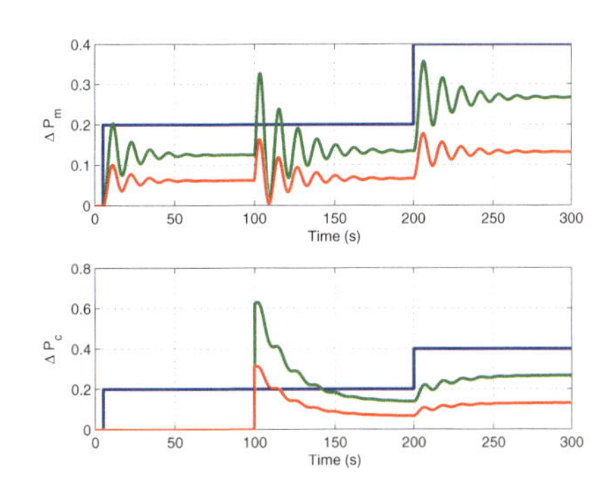

(b) Case 2: ΔP_m and ΔP_c. $R_1 = 0.05, R_2 = 0.1, k_1 = 1, k_2 = 0.5.$

Figure 3.21: Case 2 simulation results for the system of two generators serving a load.

same amount of load change since $X_1 = X_2$. After a moment, the two generators swing against each other. When one generator has power increased, the other has power decreased. This is due to the component in ΔP_{e1} is $T(\Delta\delta_1 - \Delta\delta_2)$ while in ΔP_{e2}, it is $-T(\Delta\delta_1 - \Delta\delta_2)$.

- The second observation is that the integral gain determines the power sharing after secondary frequency control. This fact is usually not mentioned in textbooks. In real-world implementation, each area has only one integrator to conduct secondary frequency control. The out-

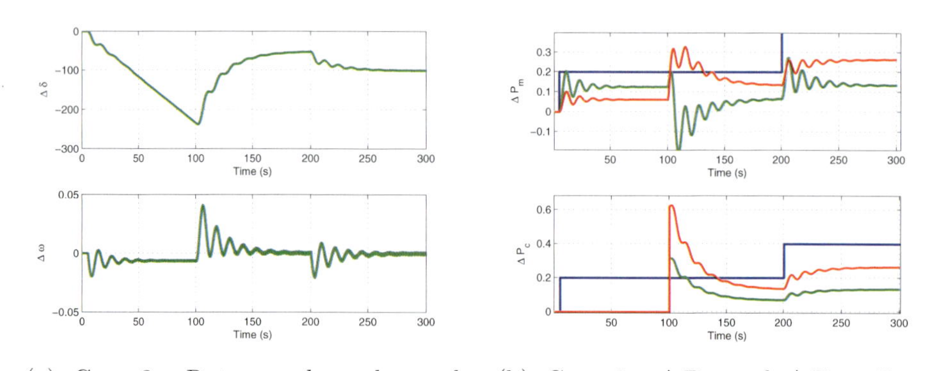

(a) Case 3: Rotor angle and speed. $R_1 = R_1 = 0.05, R_2 = 0.1, k_1 = 0.5, k_2 = 1$.

(b) Case 3: ΔP_m and ΔP_c. $R_1 = 0.05, R_2 = 0.1, k_1 = 0.5, = k_2 = 1$.

Figure 3.22: Case 3 simulation results for the system of two generators serving a load.

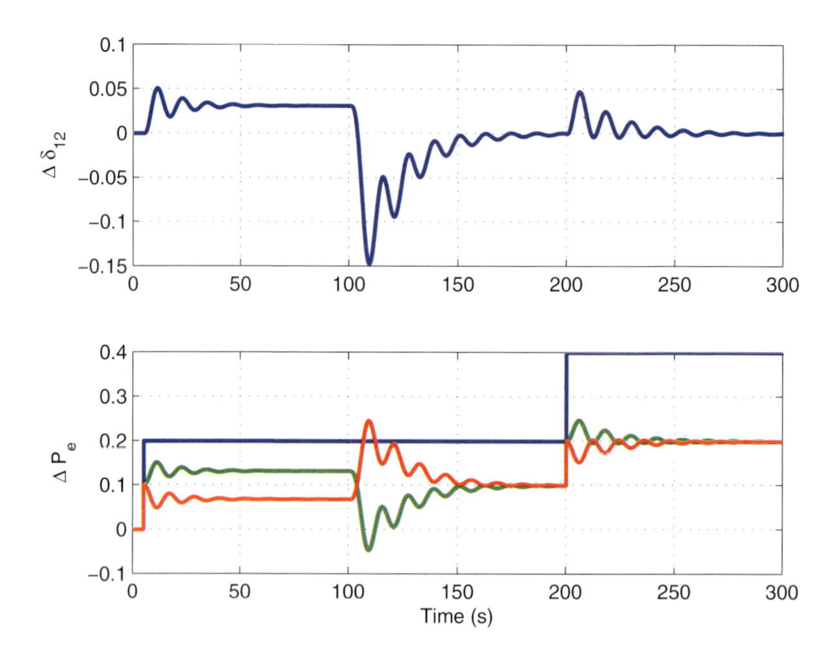

Figure 3.23: Case 1 rotor angle difference and electric power outputs for the system of two generators serving a load.

put of the integrator is then allocated to each generator through a participation factor. The effect is the same as having a different integral control gain. We can tune the gains of the secondary frequency control to let expensive generators share less and let cheap generators share more. The theoretic foundation of such sharing can be obtained through iterative solving of the Lagrangian relaxed economic dispatch problem. For more details, check the author's recent publication Miao and Fan (2017).

A few simulation tricks Note that in simulation figures, the frequencies seem not smooth. A close examination shows a lot of oscillations. This is mainly due to simulation numerical error. Simulation can be improved by limiting the simulation step size to be 0.01 seconds. This step size is applicable for electromechanical dynamics. For electromagnetic dynamics, the system bandwidth is much higher and a smaller step size will be desired.

3.5.3 Two areas connected through a tie-line

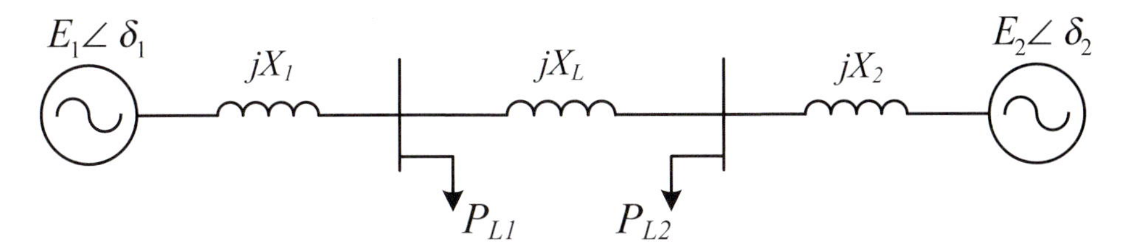

Figure 3.24: Two-area system.

Whether it is Bergen's book Bergen and Vittal (2009) or Kundur's book Kundur et al. (1994), a two-area system connected with a tie-line shown in Figure 3.24 is used for an ACE explanation. The treatment of the modeling is usually as follows. The tie-line power flow is $P_{12} = \frac{E_1E_2}{X_L}\sin(\delta_1 - \delta_2)$. And Gen 1's exporting power is $P_{L1} + P_{12}$.

In this text, using the superposition principle, an accurate expression of P_{ei} will be obtained. In addition, with certain assumptions, P_{ei} expression will be united with the expression widely adopted in other texts.

The system shown in Figure 3.24 is the superposition of three circuits as shown in Figure 3.25. The expression of P_{ei} can be found in (3.88).

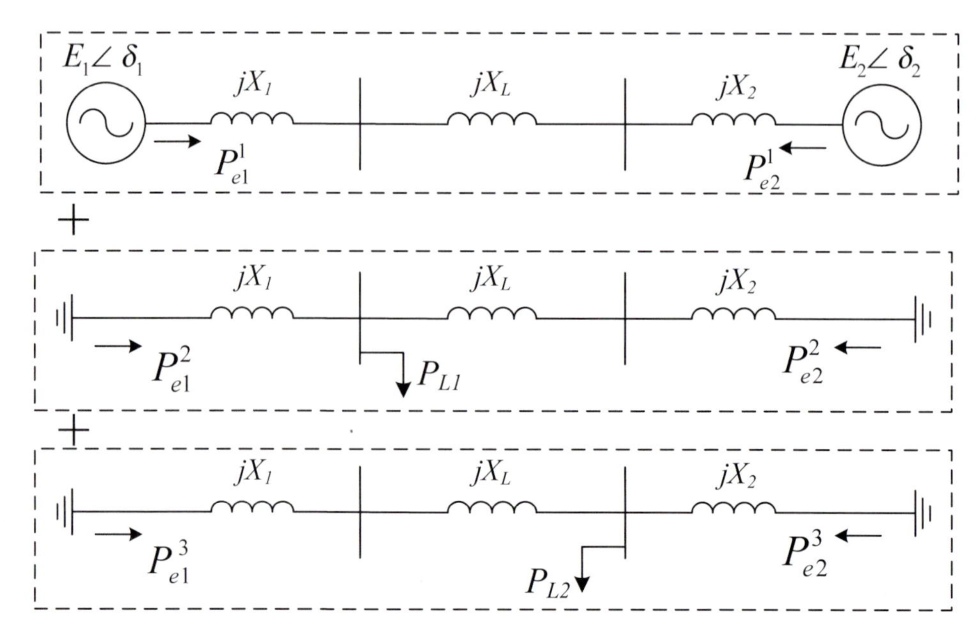

Figure 3.25: The two-area system can be viewed as the superposition of three circuits.

$$
\begin{aligned}
P_{e1} &= P_{e1}^1 + P_{e1}^2 + P_{e1}^3 \\
&= \frac{E_1E_2}{X_1+X_2+X_L}\sin(\delta_1-\delta_2) + \frac{X_2+X_L}{X_1+X_2+X_L}P_{L1} + \frac{X_1}{X_1+X_2+X_L}P_{L2} \\
P_{e2} &= P_{e2}^1 + P_{e2}^2 + P_{e2}^3 \\
&= -\frac{E_1E_2}{X_1+X_2+X_L}\sin(\delta_1-\delta_2) + \frac{X_2}{X_1+X_2+X_L}P_{L1} + \frac{X_1+X_L}{X_1+X_2+X_L}P_{L2}
\end{aligned}
$$

If we assume that compared to the line reactance, the other two reactances due to synchronizing reactance and transformer reactance can be ignored, then we have the following.

$$
\begin{aligned}
P_{e1} &= \frac{E_1E_2}{X}\sin(\delta_1-\delta_2) + P_{L1} \\
P_{e2} &= -\frac{E_1E_2}{X}\sin(\delta_1-\delta_2) + P_{L2}
\end{aligned}
\tag{3.88}
$$

where X is the total reactance of the radial system. The linearized model of P_{ei} now becomes:

$$
\begin{aligned}
\Delta P_{e1} &= T(\Delta\delta_1-\Delta\delta_2) + \Delta P_{L1} \\
\Delta P_{e2} &= -T(\Delta\delta_1-\Delta\delta_2) + \Delta P_{L2}
\end{aligned}
\tag{3.89}
$$

where $T = \frac{E_1E_2}{X}\cos(\delta_1^0-\delta_2^0)$.

Simulink blocks The Simulink blocks are shown in Figure 3.26. Note the input to the integral control is now ACE, a combination of frequency deviation and the tie-line flow deviation.

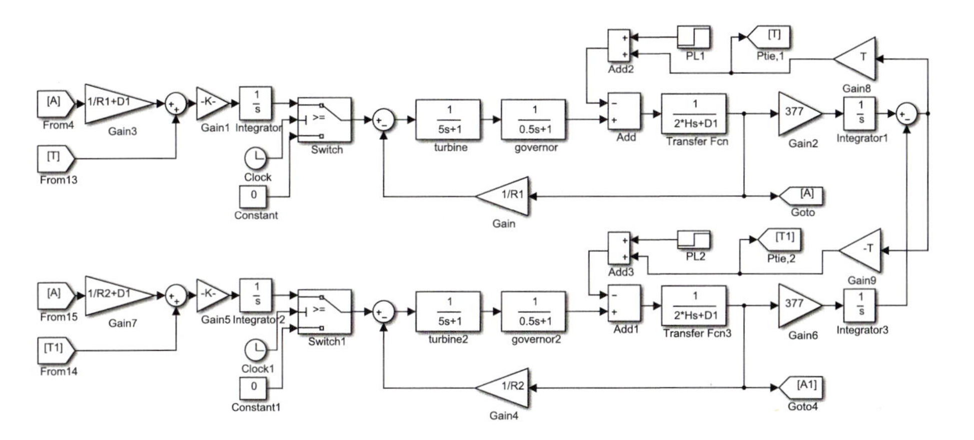

Figure 3.26: Simulink blocks for the two-area system. $H_1 = H_2 = 5, D_{1,1} = D_{2,1} = 1, T_1 = T_2 = 1, R_1 = 0.05, R_2 = 0.1$. Each integral control's gain is $1/(1/R_i + D_{1,i}) = 1/21$.

The dynamic event is as follows.

- From 0–100 seconds, the system is equipped with only droop control. At 100 seconds, Area 1's AGC is turned on. At 300 seconds, Area 2's AGC is turned on.
- At 10 seconds, Area 1 has a load increase of 0.2 pu.
- At 210 seconds, Area 2 has a load increase of 0.2 pu.

Examine the dynamic responses shown in Figure 3.27 for different time periods.

- 0–100 seconds, there is only droop control. Therefore, with a load increase in Area 1, generators in both areas respond and share the load increase. The tie-line flow now has a deviation. If we observe ACEs for two areas, we can also observe that Area 1 has an ACE $= -0.2$ pu, while Area 2's ACE is zero. ACE indicates the load change in each area. Since Area 1 has a load increase of 0.2 pu, $ACE_1 = -0.2$pu. Area 2 has no change in load, therefore, $ACE_2 = 0$.

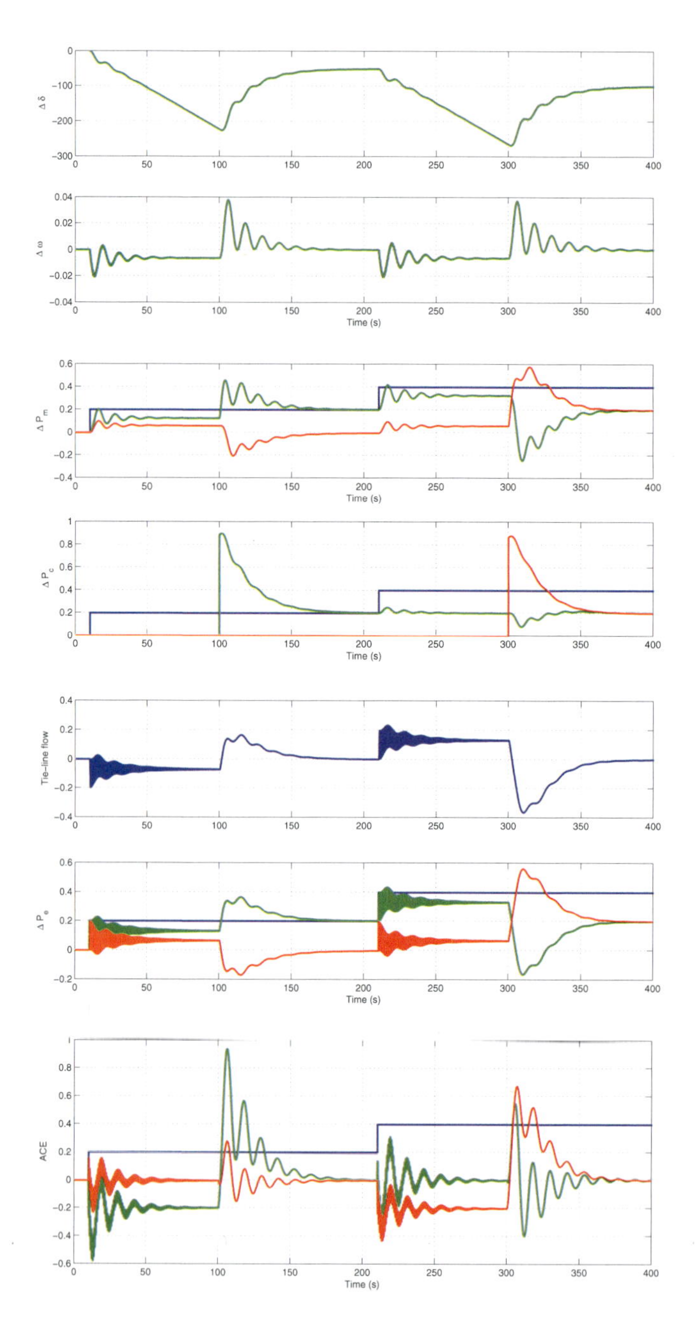

Figure 3.27: Simulation results. (1) δ. (2) ω. (3) P_m. The stair plot shows total load. (4) P_c. (5) Tie-line power flow. (6) P_e. (7) ACEs for the two areas.

- 100–210 seconds. Starting from 100 seconds, Area 1's AGC starts to work. This brings ACE_1 back to zero. The entire system's frequency is brought back to zero too. In addition, the tie-line flow is brought back to the scheduled value. For this to happen, we can see it is due to Area 1 generator's action. Its mechanical power, power order and electric power output all follow Area 1's load increase. This makes sure that the tie-line flow returns to the scheduled value. For the generator in Area 2, since Area 1 has taken care of the load increase, it will have its mechanical power, power order and electric power output all be zero. That is, Area 2 no longer shares any responsibility for the load increase.

- 210–300 seconds. Starting from 210 seconds, Area 2 has a load increase of 0.2 pu. Area 1's ACE can still be zero even if there is frequency deviation. For Area 2, since there is no AGC, the system will experience a frequency deviation. This time period is very interesting to observe. We will conduct a brief analysis for this scenario.

  At steady-state, there is ACE control in Area 1, therefore:

$$\beta_1 \Delta\omega + \Delta P_{tie,1} = 0.$$

  Note that the tie-line power comes from Gen 1's electric power with the load increase in Load 1 subtracted: $\Delta P_{tie,1} = \Delta P_{e1} - \Delta P_{L1}$. Therefore, we find that

$$\Delta P_{e1} = -\beta_1 \Delta\omega + \Delta P_{L1}.$$

  For Area 2, since there is no ACE control, its electric power and the frequency deviation relationship is

$$\Delta P_{e2} = -\beta_2 \Delta\omega.$$

  Note that the total electric power should meet the load:

$$\Delta P_{e1} + \Delta P_{e2} = \Delta P_{L1} + \Delta P_{L2}.$$

  We can find the frequency deviation:

$$\Delta\omega = -\frac{\Delta P_{L2}}{\beta_1 + \beta_2}.$$

- 300 seconds afterward. Starting from 300 seconds, Area 2's AGC is enabled. Now the entire system frequency is brought back to zero. Also the tie-line flow deviation is brought back to zero. Both generators' power orders are changed to share the load increase. Each shares 0.2 pu since the final sharing should be exactly each area's load change.

AGC control improvement Note the tie-line has oscillations and the ACE signals also have fast oscillations. Indeed, for ACE signals, we'd like them to only reflect the DC values. Therefore, a low-pass filter is applied before the ACE signals are sent to the integrator. The load changes are modified to be $\Delta P_{L1} = 0.2$ pu and $\Delta P_{L2} = 0.3$ pu. The next set of simulation results in Figure 3.28 will show better ACE signals and the final power sharing as 0.2 pu and 0.3 pu.

3.6 More examples of frequency control

3.6.1 Example 1: Step response of reference power ΔP_c

This example comes from Bergen's book in the exercise section of frequency control. This example will give a clear idea on the mechanism of secondary frequency control. In a nutshell, the secondary frequency control's job is to make sure that the power order ΔP_c matches the load ΔP_L. With this task done, the system frequency will be back to nominal. Through this example, we will show how to use MATLAB's transfer function matrix to examine step responses or linear system responses.

The system is shown in Figure 3.29 and treated as a two-input one-output system, or a general multi-input multi-output (MIMO) system. We can write the s-domain frequency expression using the superposition principle: consider one input at a time and find the summation of the effects.

$$\Delta\omega(s) = \frac{1}{1 + \frac{1}{s+1}\frac{10}{1+10s}\frac{1}{R}} \left(\frac{-10}{1+10s}\Delta P_L + \frac{1}{s+1}\frac{10}{1+10s}\Delta P_c \right) \tag{3.90}$$

Note the entire system has a loop gain and two forward gains. They are

$$\begin{aligned} &\text{Loop gain}: \frac{1}{s+1}\frac{10}{1+10s}\frac{1}{R}; \\ &\Delta P_L \to \Delta\omega: \frac{-10}{1+10s}; \\ &\Delta P_c \to \Delta\omega: \frac{1}{s+1}\frac{10}{1+10s}. \end{aligned} \tag{3.91}$$

The overall input/output relationship can be expressed using a matrix/vector format:

$$\Delta\omega(s) = \underbrace{\frac{1}{1 + \frac{1}{s+1}\frac{10}{1+10s}\frac{1}{R}} \begin{bmatrix} \frac{-10}{1+10s} & \frac{1}{s+1}\frac{10}{1+10s} \end{bmatrix}}_{G(s)} \begin{bmatrix} \Delta P_L \\ \Delta P_c \end{bmatrix} \tag{3.92}$$

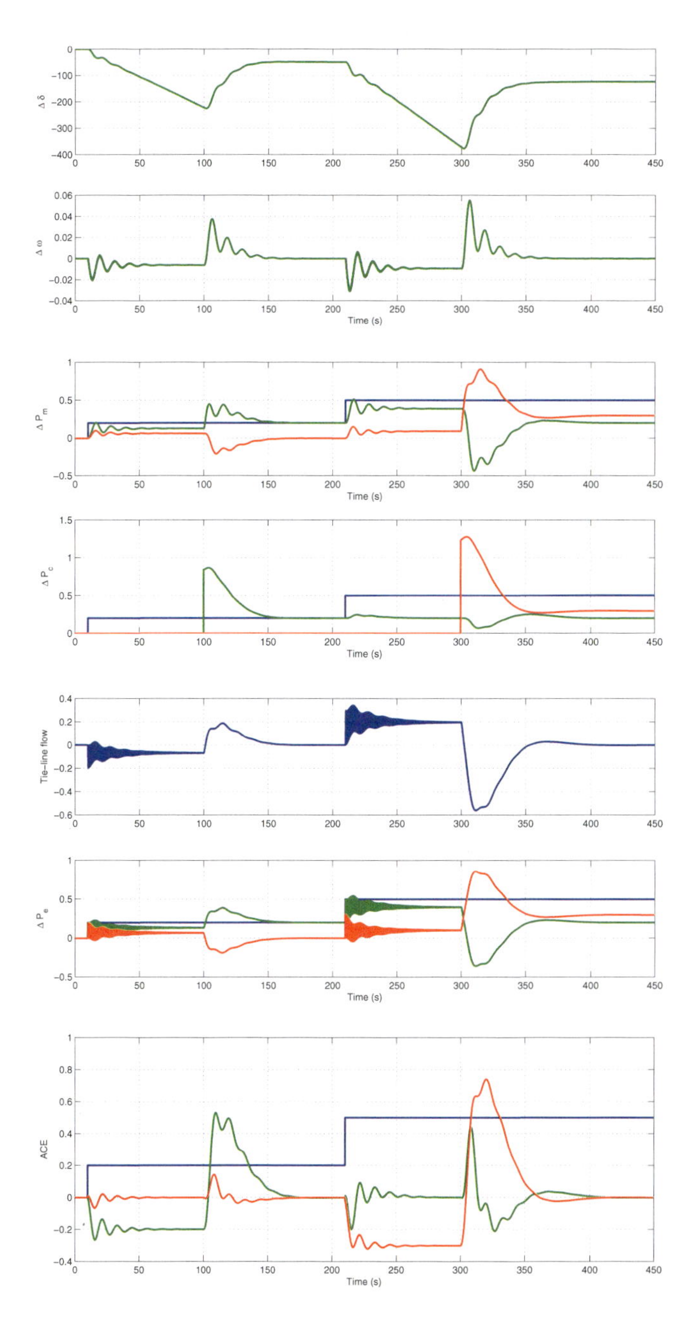

Figure 3.28: Two-area system simulation results for Case 2, ACE signals passed through a filter $1/(5s + 1)$ to the integral control.

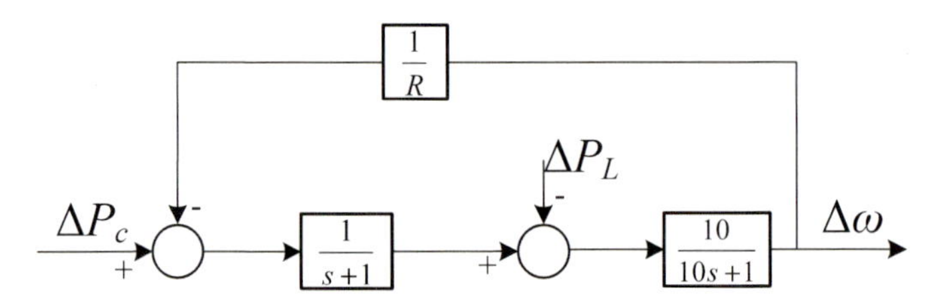

Figure 3.29: Example 1.

$G(s)$ is the transfer function matrix.

MATLAB function "*step*" examines the step responses of the system if each input signal is given a step increase.

```
s=tf('s');
Ds = 1+1/(s+1)*10/(10*s+1)*20;
G = 1/Ds*[10/(1+10*s), 1/(s+1)*10/(1+10*s)];
step(G);
```

Two step responses are given from the above code and are shown in Figure 3.30(a). The first step response shows that if ΔP_L jumps from 0 to 1, the speed $\Delta\omega$ will settle at -0.05 pu. The second step response shows that if ΔP_c jumps from 0 to 1, the speed $\Delta\omega$ will settle at 0.05 pu.

We can reason that if we apply the two step changes consecutively (first the load is increased by 1 pu, then the reference power is increased by 1 pu), the frequency will settle at the nominal. In other words, the frequency deviation will be 0. This experiment shows that the mechanism of secondary frequency control of bringing frequency back to nominal relies on making $\Delta P_c = \Delta P_L$.

The consecutive step change can be simulated using MATLAB function "lsim", which examines dynamic response for given input signals. The following MATLAB code defines an input signal with two step responses. At $t = 1$ second, the first input u_1 or ΔP_L jumps from 0 to 1. At $t = 10$ second, the second input or ΔP_c jumps from zero to 1.

```
for i=1:2000
  t(i)=0.01*(i-1);
    if (t(i)<1)  u(i,:) =zeros(1,2);
    else if(t(i)<10)
            u(i,1) = 1;  u(i,2) = 0;
        else
            u(i,1) = 1;  u(i,2) = 1;
        end
    end
```

```
end
lsim(G,u,t);
```

The simulation result is presented in Figure 3.30(b). It can be observed that with a load increase, the frequency will drop. Then with an increase in the reference power, the frequency will be brought back to the nominal.

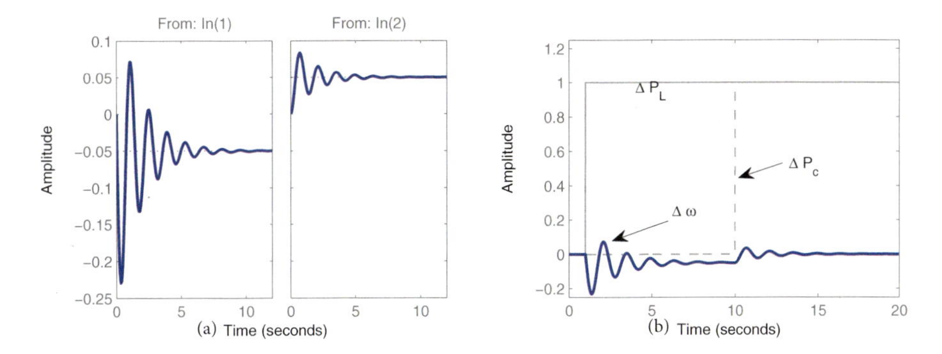

Figure 3.30: Example 1 step responses (a) and *lsim* simulation (b).

3.6.2 Example 2: Power sharing after secondary frequency control

Case 1: Why is the power sharing according to the integral control gain? The power sharing among generators with secondary frequency control should be according to the integral control gains: $k_1 : k_2 : ... : k_n$ based on the simulation study results.

This phenomenon can also be comprehended intuitively. Since the frequencies at steady-state or quasi steady-state are the same everywhere, then their integrals over a time period should be the same. Therefore, for integral control with a gain, the outputs of the integral controls are dependent on the gains. Since the integral control outputs are the power orders, therefore the generators will follow these orders and share power according to the integral gains.

The assumption that frequency everywhere is the same is the key. However, we do observe in simulation that speeds have differences. Will the differences invalidate the assumption? To better understand the issue, an accurate derivation of power sharing based on transfer function analysis and circuit analysis is given as follows.

Method 1 First, let's establish the relationship between the power order ΔP_c, ΔP_L and the rotor angles. We should have the following equations if we ignore the dynamics of the turbine governors.

$$\begin{aligned}\Delta P_{c1} - T(\Delta\delta_1 - \Delta\delta_2) &= \mathrm{PF}_1 \Delta P_L \\ \Delta P_{c2} + T(\Delta\delta_1 - \Delta\delta_2) &= (1 - \mathrm{PF}_1)\Delta P_L \end{aligned} \tag{3.93}$$

where PF_1 notates the percentage of the load power sharing of Gen 1. Gen 2's sharing is thus $1 - \mathrm{PF}_1$.

Next we have the following relationship after examining the control block diagrams.

$$\Delta P_{ci} = k_i \Delta\omega_i / s \tag{3.94}$$

$$\omega_0 \Delta\omega_i / s = \Delta\delta_i \tag{3.95}$$

Therefore

$$\Delta\delta_1 = \omega_0 \Delta P_{c1} / k_1 \tag{3.96}$$

$$\Delta\delta_2 = \omega_0 \Delta P_{c2} / k_2 \tag{3.97}$$

Substituting $\Delta\delta_i$ by the above equations in (3.93), we have:

$$\begin{aligned}\Delta P_{c1} - T\omega_0 \left(\frac{\Delta P_{c1}}{k_1} - \frac{\Delta P_{c2}}{k_2}\right) &= \mathrm{PF}_1 \Delta P_L \\ \Delta P_{c2} + T\omega_0 \left(\frac{\Delta P_{c1}}{k_1} - \frac{\Delta P_{c2}}{k_2}\right) &= (1 - \mathrm{PF}_1)\Delta P_L \end{aligned} \tag{3.98}$$

At steady-state, $\Delta P_{c2} = \Delta P_L - \Delta P_{c1}$.

$$\Delta P_{c1} - T\omega_0 \left(\frac{\Delta P_{c1}}{k_1} - \frac{\Delta P_L - \Delta P_{c1}}{k_2}\right) = \mathrm{PF}_1 \Delta P_L \tag{3.99}$$

Finally, we should have

$$\Delta P_{c1} = \frac{\mathrm{PF}_1 \Delta P_L - T\omega_0 \Delta P_L / k_2}{1 - T\omega_0(\frac{1}{k_1} + \frac{1}{k_2})} \tag{3.100}$$

$$= \frac{T\omega_0 k_1 - \mathrm{PF}_1 k_1 k_2}{(k_1 + k_2)T\omega_0 - k_1 k_2} \Delta P_L \tag{3.101}$$

Choose $T = 1$. k_1 and k_2 are in the range of 0.1–1 to have a desired bandwidth and to be separated from the primary frequency response. Note also $\omega_0 = 377$ rad/s. Therefore, $k_1 k_2 \ll T\omega_0 k_1$ and $k_1 k_2 \ll T\omega_0(k_1 + k_2)$.

With this condition, we have

$$\Delta P_c = \frac{k_1}{k_1 + k_2} \Delta P_L \tag{3.102}$$

Method 2 The second method uses the relationship of $\Delta P_{ei} = \Delta P_{ci}$ at steady-state. In addition, at any time (dynamic or steady-state)

$$\Delta P_{ci} = \left(\frac{1}{R_i} + \frac{k_i}{s}\right)\Delta\omega.$$

Therefore

$$\frac{\Delta P_{c1}}{\Delta P_{c2}} = \frac{\frac{1}{R_1} + \frac{k_1}{s}}{\frac{1}{R_2} + \frac{k_2}{s}} = \frac{k_1 + s/R_1}{k_2 + s/R_2} \tag{3.103}$$

At steady-state, the ratio becomes k_1/k_2.

Case 2: One generator is with secondary frequency control, the other is without. In that case, let's examine what is the steady-state frequency deviation and each generator's power ΔP_{ei}.

First of all, at steady-state, the input of an integral control has to be zero. Therefore, $\Delta\omega_1 = 0$. The system is an interconnected system. For an interconnected system, at steady state, frequencies everywhere are the same. Therefore, $\Delta\omega_{ss} = 0$ or the system's frequency will be brought back to nominal by Gen 1.

Second, what are the power sharing? For Gen 2, since there is no secondary frequency control, $\Delta P_{c2} = 0$. There should be no power order change. The only influencing factor for mechanical power change is from the droop control. Since the frequency deviation is zero, $\Delta P_{m2} = -\Delta\omega/R = 0$.

Therefore, Gen 1 takes care of all the load increase.

$$\Delta P_{e1} = \Delta P_L = \Delta P_{m1} - D_{1,1}\Delta\omega = \Delta P_{c1} - \Delta\omega/R_1 - D_{1,1}\Delta\omega = \Delta P_{c1} \tag{3.104}$$

The power order of Gen 1 will be increased to compensate for the load change.

3.6.3 Example 3: What if some areas have no ACE control?

Let's look at the two-area system and assume that Area 1 is with ACE control while Area 2 is without ACE. Now consider two scenarios one by one: 1) in the first scenario, area 1 load has an increase, $\Delta P_{L1} = 1$; 2) in the second scenario, area 2 has an increase, $\Delta P_{L2} = 1$. What will happen to the steady-state frequency ω_{ss}? Will it be 60 Hz? How about power sharing among the two generators in the two areas?

Scenario 1: $\Delta P_{L1} = 1$

Based on fact that the integral control's input should be zero at steady-state, Area 1's ACE signal should be zero. Therefore

$$ACE_1 = 0 = \Delta P_{12} + \beta_1 \Delta\omega = \Delta P_{e1} - \Delta P_{L1} + \beta_1 \Delta\omega. \tag{3.105}$$

From the above equation, we find:

$$\Delta P_{e1} = -\beta_1 \Delta\omega + \Delta P_{L1}. \tag{3.106}$$

Area 2 has no ACE control. Hence Gen 2's power change is due to droop and damping only:

$$\Delta P_{e2} = -\beta_2 \Delta\omega. \tag{3.107}$$

The load change will be compensated by the two generators' electric power.

$$\Delta P_{L1} = \Delta P_{e1} + \Delta P_{e2} = -(\beta_1 + \beta_2)\Delta\omega + \Delta P_{L1} \tag{3.108}$$

To make the above equation true, $\Delta\omega = 0$. Area 1 takes care of the entire load change. Area 2 does nothing.

$$\Delta P_{e1} = \Delta P_{L1} \tag{3.109}$$

$$\Delta P_{e2} = 0 \tag{3.110}$$

The system will have nominal frequency at steady-state if Area 1 has a load increase.

Scenario 2: $\Delta P_{L2} = 1$

Based on fact that the integral control's input should be zero at steady-state, Area 1's ACE signal should be zero. Therefore

$$ACE_1 = 0 = \Delta P_{12} + \beta_1 \Delta\omega = \Delta P_{e1} + \beta_1 \Delta\omega. \tag{3.111}$$

From the above equation, we find:

$$\Delta P_{e1} = -\beta_1 \Delta\omega. \tag{3.112}$$

Area 2 has no ACE control. Hence Gen 2's power change is due to droop and damping only:

$$\Delta P_{e2} = -\beta_2 \Delta\omega. \tag{3.113}$$

The load change will be compensated by the two generators' electric power.

$$\Delta P_{L2} = \Delta P_{e1} + \Delta P_{e2} = -(\beta_1 + \beta_2)\Delta\omega \tag{3.114}$$

To make that happen, $\Delta\omega = -\frac{\Delta P_{L2}}{\beta_1+\beta_2}$. Area 1 and Area 2 share based on droop and damping, or β.

$$\Delta P_{e1} = \frac{\beta_1}{\beta_1 + \beta_2}\Delta P_{L2} \tag{3.115}$$

$$\Delta P_{e2} = \frac{\beta_2}{\beta_1 + \beta_2}\Delta P_{L2} \tag{3.116}$$

The system will experience a frequency drop if Area 2 has a load increase. The entire system works as if there is no AGC control installed if one area is not equipped with ACE control yet it suffers load change.

3.6.4 Example 4: Effect of the droop in system stability

When secondary frequency control is enabled, droop seems useless. At a steady-state nominal frequency condition, droop contributes zero power. So shall we take out droop?

The answer is no. If we take out droop, the system will be unstable. Oscillations will start. Figure 3.31 shows the root loci of the system when the droop is not in place. The closed-loop system is opened at the X place. The closed-loop system will be treated as a negative feedback system. Therefore, the transfer function of the open-loop system is

$$\frac{1}{s+1}\frac{10}{10s+1}\frac{K}{s}.$$

There are three poles: 0, -1, -0.1. The root loci sketch shows that two of

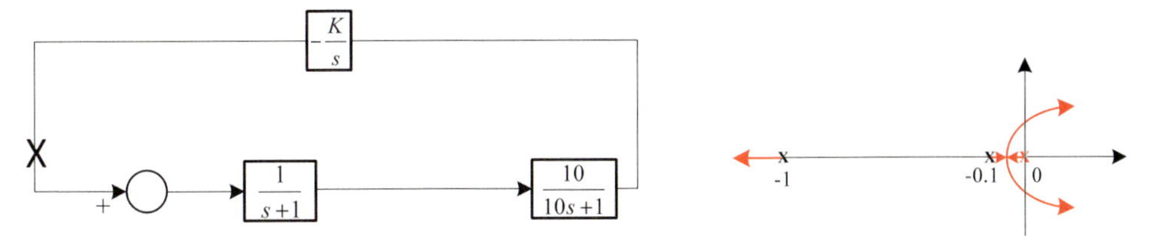

Figure 3.31: Root loci when the system has no droop control.

the closed-loop system poles will move away from 0 and -0.1 to the RHP when the gain K increases. For the above system, K should be less than 0.11 for the system to be stable.

This example shows that for a generator without droop control, in order for the system to be stable, the gain of the integral control has to be reduced dramatically. This will slow down the secondary frequency control.

The effect of droop control is explained in two steps. First, we do not consider the integral frequency control and examine the effect of droop control on the system without integral control. For the closed-loop system in Figure 3.32, the system is decoupled at X position. The open-loop system has a transfer function $\frac{1}{s+1}\frac{10}{10s+1}\frac{1}{R}$. Root loci sketch for $\frac{1}{s+1}\frac{10}{10s+1}$ is also shown in Figure 3.32. We can see that the closed-loop system with droop control will have two complex conjugated poles. If $R = 0.05$, then $1/R = 20$ and the closed-loop poles will be located at $-0.5500 \pm j4.4494$.

The closed-loop system's transfer function is $\frac{10}{\frac{10}{R}+(s+1)(10s+1)}$.

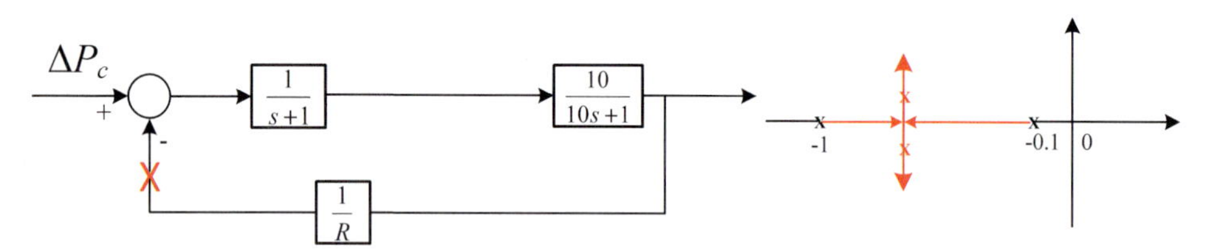

Figure 3.32: System with droop only.

Finally we consider the integral control. The system is shown in Figure 3.33. Open the system at X position. The root loci sketch is shown in Figure 3.33. For this system, if $R = 0.05$, the maximum K is 22 based on the root loci.

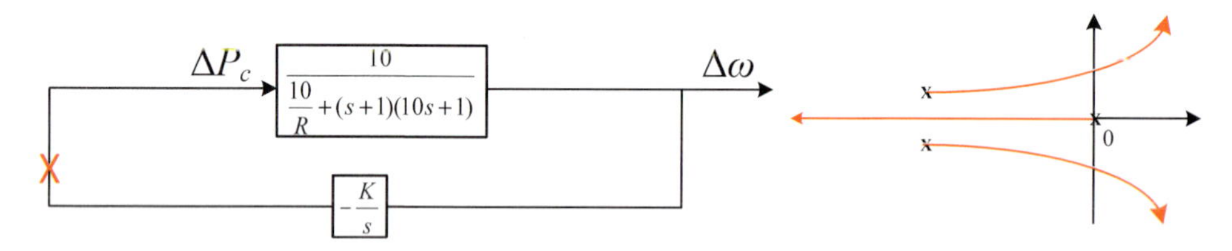

Figure 3.33: Droop control enhances system stability.

Remarks: The system with droop is more stable. The integral control gain can be much larger. Without droop, the integral control gain has to be small for the system to be stable.

Next we will examine why a large gain K for the integral control can cause instability. This can be easily shown in Figure 3.33 that a large gain K

will cause the closed-loop system poles to move to RHP. Here an alternative approach is used for the explanation. The integral control and the droop control will be aggregated as $\frac{K}{s}+\frac{1}{R}$ as shown in Figure 3.34. The effect is to introduce a zero and a pole at the original point for the open-loop system. The newly introduced zero will attract the pole at 0, while the rest of the two closed-loop system poles will be located within -0.1 and 1.0.

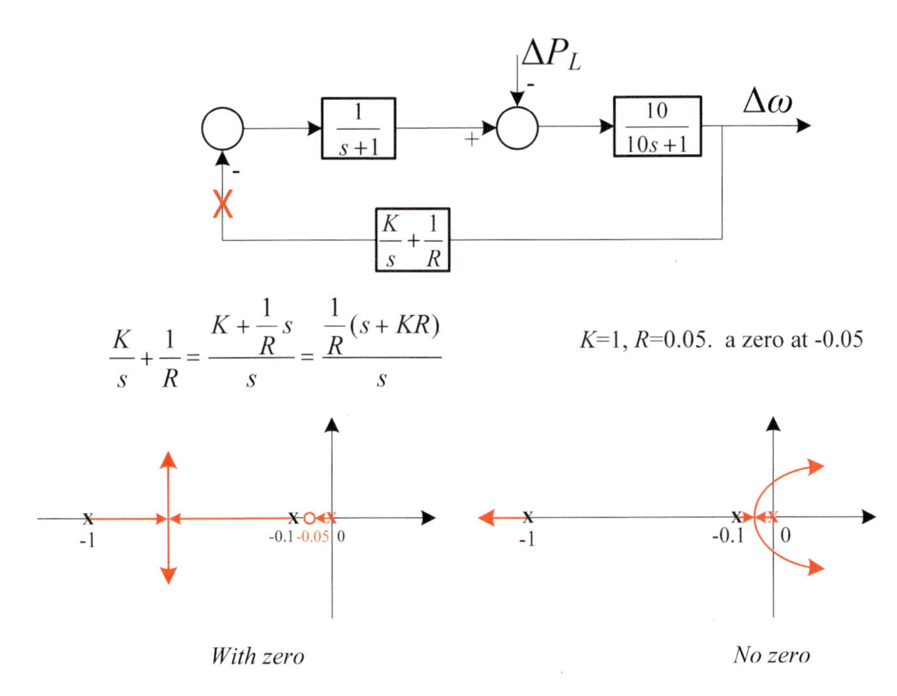

Figure 3.34: Droop control enhances system stability.

A large K causes the zero $-KR$ to move to the left. The original point of the asymptotes is related to the poles and zeros as following:

$$s_0=\frac{\sum p_i-\sum z_i}{n-m} \quad \text{center-of-gravity rule.} \tag{3.117}$$

where p_i are the poles and z_i are the zeros, n is the number of the poles and m is the number of zeros.

A large gain will make the s_0 move toward the RHP. The root loci of an example system with different integral control gain K is shown in Figure 3.35. The crosses show the positions of the closed-loop system poles when the gain of the loop gain $\frac{1}{s}\frac{10}{10s+1}\frac{s+kR}{s}$ is at 20 (since $1/R=20$). Figure 3.35 shows that a large K causes system instability. When $K=1$ and $K=10$, the system has closed-loop poles located at the LHP. When $K = 100$, the system has poles that are located in the RHP.

```
s=tf('s');
k=[1,10,100];
for i=1:3;
   P(i) = 1/s/(s+1)/(s+0.1)*(s+k(i)*0.05);
end

for i=1:3
   figure(i);
   rlocus(P(i)); hold on;
   x= rlocus(P(i), 20);
   plot(real(x), imag(x),'b+');
end
```

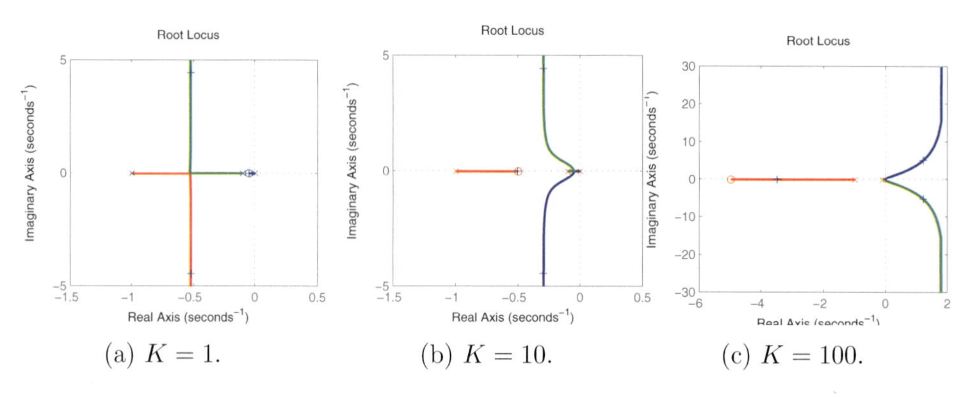

(a) $K = 1$. (b) $K = 10$. (c) $K = 100$.

Figure 3.35: Effect of K. $K = 1, K = 10, K = 100$. A large K will cause system instability.

3.6.5 Example 5: Why does long distance transmission induce more oscillations?

In simulation results shown in Figure 3.27, the two-area system connected with long-distance tie-lines shows more oscillations in tie-line power flow and electric power compared to the simulation results of the system where two generators serve a single load. We can examine the electric power expressions for these two cases. In the long-distance case:

$$P_{e1} = \frac{E_1 E_2}{X}\sin(\delta_1 - \delta_2) + P_{L1} \tag{3.118a}$$

$$P_{e2} = -\frac{E_1 E_2}{X}\sin(\delta_1 - \delta_2) + P_{L2} \tag{3.118b}$$

In the two-generator serving one load case:

$$P_{e1} = P_{e1}^1 + P_{e1}^2 = \frac{E_1 E_2}{X_1 + X_2}\sin(\delta_1 - \delta_2) + \frac{1}{2}P_L \tag{3.119a}$$

$$P_{e2} = P_{e2}^1 + P_{e2}^2 = \frac{E_1 E_2}{X_1 + X_2}\sin(\delta_2 - \delta_1) + \frac{1}{2}P_L \tag{3.119b}$$

For the two-generator serving a load case, examine the dynamics of $\Delta(\delta_1 - \delta_2)$ while assuming $\Delta P_m = 0$:

$$\begin{aligned}
&\frac{d}{dt}\Delta(\delta_1 - \delta_2) = \omega_0(\Delta\omega_1 - \Delta\omega_2)\\
&\frac{d}{dt}\Delta(\omega_1 - \omega_2) = -\left(\frac{T}{2H_1} + \frac{T}{2H_2}\right)\Delta\delta_{12} - \left(\frac{\Delta P_L}{4H_1} - \frac{\Delta P_L}{4H_2}\right) - \frac{D_{1,1}\Delta\omega_1}{2H_1} - \frac{D_{1,2}\Delta\omega_2}{2H_2}
\end{aligned} \tag{3.120}$$

where $T = \frac{E_1 E_2}{X_1 + X_2}\cos(\delta_2^0 - \delta_1^0)$.

If $H_1 = H_2$, then for the rotor angle difference dynamics, the effect of ΔP_L can be ignored. On the other hand, if H_1 does not equal H_2, ΔP_L is enforced. This will cause oscillations. The effect is sort of like a long-distance connection. That is the reason for two generators with very different sizes, we do not put them on the same bus to serve a load.

For the two-area system with a long-distance tie-line, Load 1 and Load 2 effects are always enforced if $\Delta P_{L1} \neq \Delta P_{L2}$:

$$\begin{aligned}
&\frac{d}{dt}\Delta(\delta_1 - \delta_2) = \omega_0(\Delta\omega_1 - \Delta\omega_2)\\
&\frac{d}{dt}\Delta(\omega_1 - \omega_2) = -\left(\frac{T}{2H_1} + \frac{T}{2H_2}\right)\Delta\delta_{12} - \left(\frac{\Delta P_{L1}}{2H_1} - \frac{\Delta P_{L2}}{2H_2}\right) - \frac{D_{1,1}\Delta\omega_1}{2H_1} - \frac{D_{1,2}\Delta\omega_2}{2H_2}
\end{aligned} \tag{3.121}$$

where $T = \frac{E_1 E_2}{X}\cos(\delta_2^0 - \delta_1^0)$.

Therefore, we observe more oscillations in generators' electric power for the two-area with tie-line case (Figure 3.27) compared with that for the two-generator serving a load case (Figure 3.23). Note that in both cases, we assume that the two generators in the system have the same inertia H, and damping D_1.

Exercises

1. Use parameters in Example 1 Figure 3.29. For a single generator load serving system, derive the linear system model and build the model in MATLAB/Simulink. Find the droop to make $\Delta\omega = -0.2$ for $\Delta P_L = 0.1$.

- Find the bandwidth of the system with only primary frequency control.
- Provide the dynamic simulation of the system frequency due to a step response of load increase 0.1.
- Specify ΔP_c to bring $\Delta\omega$ back to zero.
- Design the secondary frequency control to bring the system frequency back to nominal when load varies. Choose a gain of the integration block to make the frequency return to nominal in less than 100 seconds. Find the value of ΔP_c at steady-state and see if this value matches your previous calculation. Provide the dynamic responses of $\Delta\omega$ and ΔP_c.
- Increasing the gain of the integration block will lead to instability. Find the marginal value of the gain. Designate an analysis procedure to analytically find the marginal gain. Hint: you can use root loci method.

2. Use the parameters in Example 3. In MATLAB/Simulink, build the linearized model for a two-area interconnection system. Each area consists of a generator with a load. A load change of 100 MW occurs in Area 1.

- Demonstrate the ACE steady-state values, steady-state frequency in Hz, change in tie-line flow in MW for each area through time-domain simulation for the system with primary frequency control only. Observe if the values match the calculation.
- Design secondary frequency control for each area and demonstrate that the frequency and tie-line power flow will go back to nominal. Please present simulation results.

Chapter 4

Synchronous Generator Models

In this chapter, steady-state models and dynamic models of a synchronous generator will be presented. For a round-rotor generator, the steady-state model can be represented by a voltage source behind a synchronous reactance. For a salient generator, such a simple circuit representation is not possible. Instead, a phasor diagram is used more popularly to represent the relationship among terminal voltage, internal voltage, and current.

The electromagnetic dynamics of a synchronous generator are related to the voltage and flux linkage of a circuit, which can be expressed by Faraday's Law. The dynamic models of a synchronous generator are expressed in the rotor reference frame, or dq-reference frame. This reference frame is rotating at the nominal speed at steady-state. Modeling a synchronous generator in the dq-reference frame is a very important technique. The conversion of variables from the abc frame to the dq frame is the well-known Park's transformation. Park's 1929 paper Park (1929) was voted the second most influential paper in power engineering in the 20th century in 2000 Heydt et al. (2000), second only to the first influential paper by Fortescue in 1918 on symmetrical component theory Fortescue (1918).

Park's transformation has been applied in the analysis of synchronous or asynchronous machines. Through the transformation, stator variables are expressed in the rotor reference frame. The main advantage is that the related linear differential equations with time varying inductances become linear differential equations with time *invariant* coefficients.

The treatment of Park's transformation in classic texts, e.g., Bergen and Vittal (2009), Krause (1986), relies on a transformation matrix $\mathbf{P}$ in real domain ($\mathbf{i}_{dq0} = \mathbf{P}\mathbf{i}_{abc}$). In this text, the space vector concept will be introduced and applied in reference frame conversion. The introduction of space vector concept will lead to a straightforward procedure on steady-state and dynamic model derivation.

4.1 Generator steady-state circuit model

The steady-state circuit model of a generator is derived based on the superposition principle. First, we consider rotor flux only. Then we consider stator current effect only (armature reaction). The two effects will then be combined to derive the circuit model, and further phasor diagram and power expressions.

4.1.1 Internal voltage due to the rotor excitation current

A cross section of a two-pole synchronous generator with a salient rotor is shown in Figure 4.1.

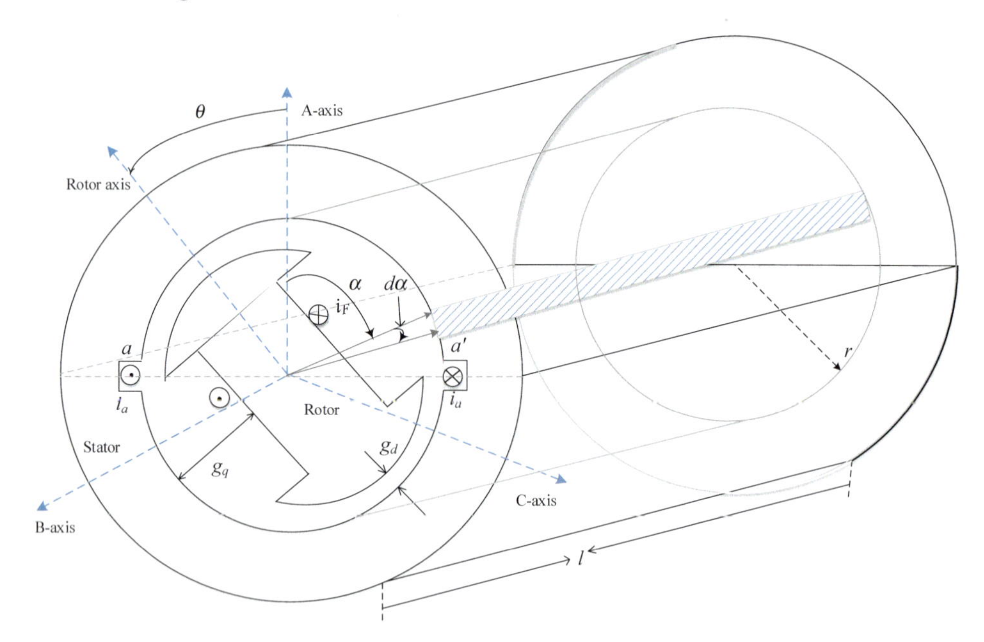

Figure 4.1: Synchronous generator cross section and the Gaussian surface. θ is the rotor position relative to the reference axis.

The rotor circuit is excited by a DC excitation voltage v_F and the current i_F. The rotor is rotating at ω speed. This DC current along with the motion will produce flux as a traveling waveform in the air gap. If we use Ampere's Law, we should be able to find the field strength and flux density in the air gap.

$$\oint_{\Gamma} H dl = N_F i_F \tag{4.1}$$

where H is the magnetic field strength, Γ is the flux link path, and N_F is the number of rotor windings. The air gap magnetic permeability is much less than that in the rotor and stator. Therefore, if we separate the path into two portions (air gap Γ_1 and non air gap Γ_2), we have:

$$\begin{aligned} N_F i_F &= \int_{\Gamma_1} \frac{B}{\mu_0} dl + \int_{\Gamma_2} \frac{B}{\mu} dl \\ &\approx \int_{\Gamma_1} \frac{B}{\mu_0} dl \qquad \text{since } \mu \gg \mu_0 \\ &= 2 g_d \frac{B}{\mu_0} \end{aligned} \tag{4.2}$$

where B is the flux density, g_d is the air gap distance at the rotor flux line. For a round rotor, the air gap distance is uniform. For a salient rotor, the air gap distance is not uniform. However, even for a salient rotor, g_d is a fixed distance since it is the air gap distance at the rotor flux line.

The above relationship has an assumption that the magnetic field is linear, therefore,

$$\mu H = B. \tag{4.3}$$

The magnitude of the flux density in the air gap can be found as

$$B = \frac{\mu_0}{2 g_d} N_F i_F.$$

In the air gap, the flux density will be the same and have the same direction (going out from the rotor from $\frac{-\pi}{2}$ to $\frac{\pi}{2}$ based on the rotor position, going into the rotor for the rest π position. See Figure 4.2). We will define going out as the positive direction for the flux lines while going in as the negative direction.

The amplitude of the fundamental waveform for a square waveform with a magnitude of 1 is $\frac{4}{\pi}$ (see Figure 4.2). Therefore the amplitude of the fundamental component of the flux density is

$$\hat{B} = \frac{4}{\pi} \frac{\mu_0}{2 g_d} N_F i_F. \tag{4.4}$$

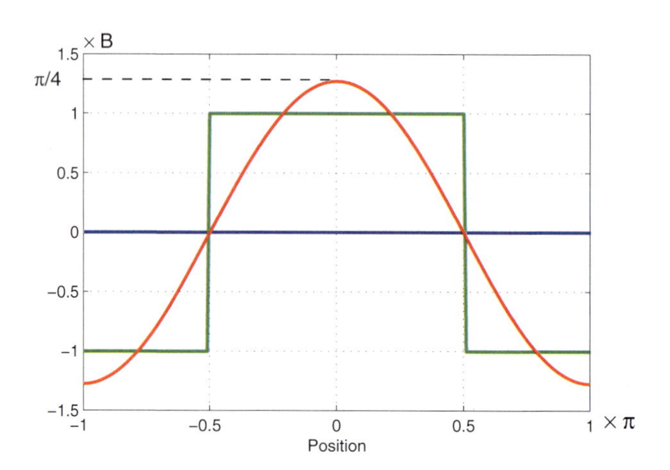

Figure 4.2: Flux density in the air gap and the fundamental waveform of the flux density. The position is relative to the rotor axis.

We now present the second assumption. Machines are designed to have a sinusoidal flux distribution in the air gap. Therefore, in the air gap, at a random position notated by α angle from the reference axis, the flux density is expressed as

$$B_F(\alpha) = \hat{B}\cos(\alpha - \theta) = \frac{4}{\pi}\frac{\mu_0}{2g_d}N_F i_F \cos(\alpha - \theta), \tag{4.5}$$

where θ is the rotor flux position or rotor position. Subscript F is used to notate rotor flux since in the next sections, other fluxes will be introduced.

We now consider the stator's phase A coil and study the flux linkage linked $\lambda_{aa'}$ due to the rotor flux B_F. To find $\lambda_{aa'}$, we have to compute the entire flux that is encompassed by coil aa' in the air gap Gaussian space. Since the density is different everywhere, integration is used.

First, check a trip of Gaussian surface that corresponds to a small angle $d\alpha$. The surface area is $rld\alpha$. The corresponding flux density is $\hat{B}\cos(\alpha-\theta)$. Next we find the integration from $-\pi/2$ to $\pi/2$.

$$\begin{aligned} \phi_{aa'} &= \int_{-\pi/2}^{\pi/2} \hat{B}\cos(\alpha - \theta) rld\alpha \\ &= 2rl\hat{B}\cos\theta \end{aligned} \tag{4.6}$$

The flux linkage linked to aa' is $N\phi_{aa'}$ with N as the number of windings of phase a.

$$\begin{aligned}\lambda_{aa'} &= 2Nrl\hat{B}\cos\theta \\ &= \underbrace{2Nrl\frac{4}{\pi}\frac{\mu_0}{2g_d}N_F}_{M_F} i_F\cos\theta = M_F i_F \cos\theta\end{aligned} \tag{4.7}$$

where M_F is defined as above and is called mutual inductance.

Similarly, if we want to find out the flux linkages linked to coils bb' and cc', then the integration should be in the range of $[\frac{2\pi}{3}-\frac{\pi}{2}, \frac{2\pi}{3}+\frac{\pi}{2}]$, and $[\frac{4\pi}{3}-\frac{\pi}{2}, \frac{4\pi}{3}+\frac{\pi}{2}]$ based on the allocation of stator coils. Note that stator coils are allocated so that phase bb' will have a reference axis ahead of that of phase aa' by 120 degree, phase cc' ahead of phase bb' 120 degree.

$$\begin{aligned}\lambda_{bb'} &= N\int_{\frac{2\pi}{3}-\frac{\pi}{2}}^{\frac{2\pi}{3}+\frac{\pi}{2}} \hat{B}\cos(\alpha-\theta)rld\alpha \\ &= 2Nrl\hat{B}\cos\left(\theta-\frac{2\pi}{3}\right)\end{aligned} \tag{4.8}$$

$$\begin{aligned}\lambda_{cc'} &= N\int_{\frac{4\pi}{3}-\frac{\pi}{2}}^{\frac{4\pi}{3}+\frac{\pi}{2}} \hat{B}\cos(\alpha-\theta)rld\alpha \\ &= 2Nrl\hat{B}\cos\left(\theta+\frac{2\pi}{3}\right)\end{aligned} \tag{4.9}$$

Based on Faraday's Law, a varying flux linkage induces EMF or voltage, e.g., $e_{a'a} = \frac{d\lambda_{aa'}}{dt}$. In addition, instead of using $e_{a'a}$, we will use $e_{aa'}$ as the voltage by treating the generator as a voltage source with currents flowing out of the generator. Hence we have

$$e_{aa'} = -\frac{d\lambda_{aa'}}{dt} = \dot{\theta}M_F i_F \sin\theta = \omega M_F i_F \cos\left(\theta-\frac{\pi}{2}\right). \tag{4.10}$$

Considering the nominal condition when the speed is nominal ω_0 and $\theta = \omega_0 t + \theta_0$, we have the internal voltage $e_{aa'}$ and its corresponding phasor $\overline{E}_a$ as

$$\begin{aligned}e_{aa'} &= -\frac{d\lambda_{aa'}}{dt} = \dot{\theta}M_F i_F \sin\theta = \omega M_F i_F \cos\left(\omega_0 t + \theta_0 - \frac{\pi}{2}\right), \\ \overline{E}_a &= \frac{\omega M_F i_F}{\sqrt{2}} e^{j(\theta_0-\frac{\pi}{2})}.\end{aligned} \tag{4.11}$$

Define $\delta = \theta_0 - \frac{\pi}{2}$, then

$$\overline{E}_a = \frac{\omega M_F i_F}{\sqrt{2}} e^{j\delta} \tag{4.12}$$

θ_0 is the initial position of the rotor axis (d-axis) relative to the reference axis (static) and δ is the initial position of the quadratic-axis (q-axis) relative to the reference axis.

4.1.2 Armature reaction of a round rotor generator

In this subsection, the rotor flux is not in the picture. We will only consider three-phase stator currents i_a, i_b, i_c and their combined effect in generating a flux and an EMF.

The currents are balanced three-phase currents.

$$\begin{aligned} i_a &= I_m \cos(\theta_a) \\ i_b &= I_m \cos\left(\theta_a - \frac{2\pi}{3}\right) \\ i_c &= I_m \cos\left(\theta_a + \frac{2\pi}{3}\right) \end{aligned} \tag{4.13}$$

Assume the air gap is uniform. For phase a current i_a, using the same technique adopted in flux density calculation due to rotor current i_F, we can compute the flux density anywhere in the air gap. Further, we can extend the flux density expression due to i_b and i_c.

$$\begin{aligned} B_a(\alpha) &= \frac{4}{\pi}\frac{\mu_0}{2g} N i_a \cos\alpha \\ B_b(\alpha) &= \frac{4}{\pi}\frac{\mu_0}{2g} N i_b \cos\left(\alpha - \frac{2\pi}{3}\right) \\ B_c(\alpha) &= \frac{4}{\pi}\frac{\mu_0}{2g} N i_c \cos\left(\alpha + \frac{2\pi}{3}\right) \end{aligned} \tag{4.14}$$

where α is a place in the air gap relative to the reference a-axis.

The above expression shows that in the air gap, B_a will be maximum or minimum at zero degree (a-axis), while B_b will be maximum or minimum at 120^0 (b-axis), while B_c will be maximum or minimum at -120^0 (c-axis). Currents are time varying. Therefore, the magnitude of the flux density is also time varying. The combined flux density is

$$\begin{aligned} B_{ar}(\alpha) &= B_a(\alpha) + B_b(\alpha) + B_c(\alpha) \\ &= \frac{4}{\pi}\frac{\mu_0}{2g} N I_m \left(\cos\theta_a \cos\alpha + \cos\left(\theta_a - \frac{2\pi}{3}\right) \cos\left(\alpha - \frac{2\pi}{3}\right) \right. \\ &\left. + \cos\left(\theta_a + \frac{2\pi}{3}\right) \cos\left(\alpha + \frac{2\pi}{3}\right) \right) \\ &= \frac{4}{\pi}\frac{\mu_0}{2g} N \frac{3}{2} I_m \cos(\alpha - \theta_a) \end{aligned} \tag{4.15}$$

Subscript ar notates armature reaction.

If we compare this expression with the flux density generated by i_F shown as below

$$B_F(\alpha) = \frac{4}{\pi}\frac{\mu_0}{2g_d} N_F i_F \cos(\alpha - \theta), \tag{4.16}$$

we may realize that the effect of three-phase balanced stator currents (i_a, i_b, i_c) in a uniform air gap of a round-rotor generator) is the same as a DC rotor current $\frac{3}{2}I_m$ on a rotor with a rotating speed the same as the electric frequency.

Remarks: Rotating magnetic field is the most important concept in ac machine. It can be formed by a DC rotor current with the rotor rotating or three-phase balanced stator currents with static stators.

For B_{ar}, using the similar technique in the previous subsection to find flux linkage linked to stator coil aa', we have:

$$\begin{aligned} \lambda_{ar} &= \underbrace{2Nrl\frac{4}{\pi}\frac{\mu_0}{2g} N \frac{3}{2}}_{L_{s1}} I_m \cos(\theta_a) \\ &= L_{s1} i_a \end{aligned} \tag{4.17}$$

The induced EMF v_{ar} can be expressed as follows:

$$v_{ar} = -\frac{d\lambda_{ar}}{dt} = -L_{s1}\frac{di_a}{dt}. \tag{4.18}$$

4.1.3 Round-rotor generator circuit, phasor diagram, power and torque

Adding the rotor flux and armature reaction together, we can find the total flux linkage linked to aa' due to the air gap flux:

$$\lambda_{ag} = \underbrace{M_F i_F \cos\theta}_{\lambda_{aa'}} + \underbrace{L_{s1} i_a}_{\lambda_{ar}}. \tag{4.19}$$

where $\lambda_{aa'}$ is the flux linkage due to the rotor current and λ_{ar} is the flux linkage due to the stator currents or armature reaction.

The corresponding air gap voltage v_{ag} is

$$v_{ag} = -\frac{d\lambda_{ag}}{dt} = \omega M_F i_F \sin\theta - L_{s1}\frac{di_a}{dt} \tag{4.20}$$

If we consider the nominal condition and steady-state expression, then the flux linkages are expressed as follows.

$$\begin{aligned} \lambda_{aa'}(t) &= M_F i_F \cos(\omega_0 t + \delta + \pi/2) \\ \lambda_{ar}(t) &= L_{s1} I_m \cos(\omega_0 t + \theta_{a0}) \\ \lambda_{ag}(t) &= \lambda_{aa}(t) + \lambda_{ar}(t) \end{aligned} \tag{4.21}$$

The voltage induced by the total air gap flux linkage is as follows.

$$\begin{aligned} v_{ag}(t) &= \omega_0 M_F i_F \sin\theta + \omega_0 L_{s1} I_m \sin\theta_a \\ &= \omega_0 M_F i_F \sin(\omega_0 t + \theta_0) + \omega_0 L_{s1} I_m \sin(\omega_0 t + \theta_{a0}) \\ &= \omega_0 M_F i_F \cos(\omega_0 t + \theta_0 - \pi/2) + \omega_0 L_{s1} I_m \cos(\omega_0 t + \theta_{a0} - \pi/2). \end{aligned} \tag{4.22}$$

The phasor relation for the flux linkages is expressed as:

$$\begin{aligned} \overline{\lambda}_{ag} = \overline{\lambda}_{aa'} + \overline{\lambda}_{ar} &= \frac{M_F i_F}{\sqrt{2}} e^{j(\delta+\pi/2)} + L_{s1}\overline{I}_a \\ &= j\frac{\overline{E}_a}{\omega_0} + L_{s1}\overline{I}_a. \end{aligned} \tag{4.23}$$

The phasor relationship for the voltage and currents is expressed as:

$$\begin{aligned} \overline{V}_{ag} &= \frac{\omega_0 M_F i_F}{\sqrt{2}} e^{j(\theta_0-\pi/2)} + X_{s1}\frac{I_m}{\sqrt{2}} e^{j(\theta_a-\pi/2)} \\ &= E_a e^{j\delta} - jX_{s1}\frac{I_m}{\sqrt{2}} e^{j\theta_a} \\ &= \overline{E}_a - jX_{s1}\overline{I}_a. \end{aligned} \tag{4.24}$$

where $X_{s1} = \omega_0 L_{s1}$.

If we consider the stator resistance and leakage reactance, then we have the following expression.

$$\overline{E}_a = \overline{V}_a + (r + jX_{ls} + jX_{s1})\overline{I}_a = \overline{V}_a + (r + jX_s)\overline{I}_a \tag{4.25}$$

where X_s is called synchronous reactance.

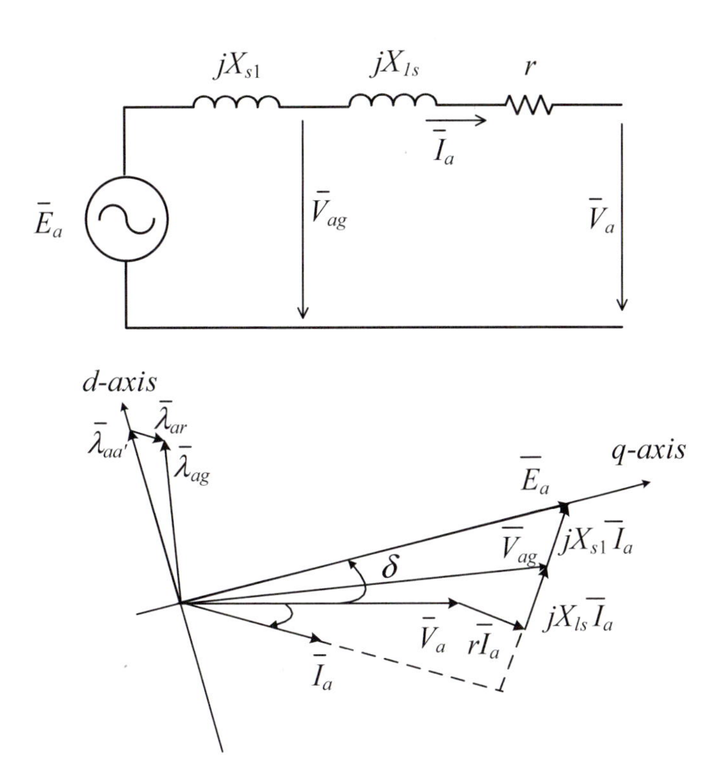

Figure 4.3: Round-rotor generator circuit model and phasor diagram.

The circuit model and phasor diagram of a round-rotor generator is shown in Figure 4.3. We start to introduce dq axes. The dq axes have been briefly mentioned in Chapter 3 and illustrated in Fig 3.1. The rotor axis shown in Figure 4.1 is called direct axis or d-axis. The quadrature axis or q-axis lags the d-axis by 90 degrees. In the phasor diagram, we also use q-axis to notate the direction of the internal voltage and use d-axis to notate the rotor flux phasor's direction.

Given the terminal voltage and current phasors, we can find the internal voltage E_a and its phase angle δ. The active power delivered by the generator can be easily found with r ignored.

$$P_a = \frac{E_a V_a}{X_s} \sin \delta \tag{4.26}$$

if we assume that the phase angle of the terminal voltage is zero $v_a(t) = \sqrt{2}V_a \cos(\omega_0 t)$. Then δ is the angle difference between the two voltage phasors $\overline{E}_a$ and $\overline{V}_a$.

4.1.4 Lentz's Law example

Lentz's Law states that if an original flux induces EMF and this EMF may generate related current, then the current will generate a flux that will weaken the original flux. This is best explained by the terminal bus short circuit example.

When a generator's terminal bus is short circuited, if we ignore the resistance, we will have the current phasor, armature flux linkage phasor, and rotor flux phasor all referring to the q-axis as

$$\overline{E}_a = E_a, \tag{4.27}$$

$$\overline{\lambda}_{aa'} = j\frac{E_a}{\omega}, \tag{4.28}$$

$$\overline{I}_a = \frac{E_a}{jX_s} = -j\frac{E_a}{X_s}, \tag{4.29}$$

$$\overline{\lambda}_{ar} = L_s\overline{I}_a = -jL_s\frac{E_a}{X_s} = -j\frac{E_a}{\omega}. \tag{4.30}$$

Therefore, the armature flux cancels the rotor flux. The total air gap flux is zero.

4.2 Space vector concept

In this section, we start to examine the concept of space vector, which has been widely used in ac machine and power electronics. As readers can sense, this concept comes from the rotating magnetic field. Based on the analysis conducted in the previous section, we have two important findings.

1. A rotating magnetomotive force (MMF) and further a rotating magnetic field are formed due to a constant DC excitation current i_F on the rotor. This constant current will produce a sinusoidal magnetic field in the air gap with constant magnitude. In addition, the rotor is rotating at speed ω. Hence this magnetic field is rotating, or, a rotating magnetic field with constant magnitude.

2. Balanced three-phase stator currents can also form a rotating MMF and further a rotating magnetic field. If the electric frequency is ω, the rotating magnetic field is rotating with a speed at ω.

Examine the physics of MMF: at any place in the air gap (notated as α angle from the reference axis), the MMF is expressed as

$$F_a(\alpha) = Ni_a \cos\alpha$$
$$F_b(\alpha) = Ni_b \cos\left(\alpha - \frac{2\pi}{3}\right)$$
$$F_c(\alpha) = Ni_c \cos\left(\alpha + \frac{2\pi}{3}\right)$$

where α is the general angle in the air gap referring to the a-axis, F is the MMF.

We can see from the above equations that F_a is maximum when $\alpha = 0$. Accordingly, F_b is maximum when $\alpha = \frac{2\pi}{3}$; F_c is maximum when $\alpha = \frac{4\pi}{3}$. Consider the currents i_a, i_b, and i_c as balanced three-phase.

$$\begin{aligned} i_a(t) &= I_m \cos\theta_a = I_m \cos(\omega_e t + \theta_a) \\ i_b(t) &= I_m \cos(\theta_a - \frac{2\pi}{3}) = I_m \cos\left(\omega_e t + \theta_a - \frac{2\pi}{3}\right) \\ i_c(t) &= I_m \cos(\theta_a + \frac{2\pi}{3}) = I_m \cos\left(\omega_e t + \theta_a + \frac{2\pi}{3}\right) \end{aligned} \tag{4.31}$$

where ω_e notates the electricity frequency.

Then we have

$$\begin{aligned} F(\alpha, t) = NI_m\Bigg[& \cos(\omega_e t + \theta_a)\cos\alpha + \cos\left(\omega_e t + \theta_a - \frac{2\pi}{3}\right)\cos\left(\alpha - \frac{2\pi}{3}\right) \\ & + \cos\left(\omega_e t + \theta_a + \frac{2\pi}{3}\right)\cos\left(\alpha + \frac{2\pi}{3}\right)\Bigg] \\ = & \frac{3}{2}NI_m \cos\left(\alpha - \omega_e t - \theta_a\right) \end{aligned} \tag{4.32}$$

For the above MMF, if we only consider its maximum in the 2D dimension of the air gap, then we find that

$$\hat{F}(t) = \frac{3}{2}NI_m \tag{4.33}$$

when $\alpha = \omega_e t + \theta_a$.

We now introduce a phasor (or a space vector) to notate the magnitude and the angle of the MMF as:

$$\vec{F}(t) = \hat{F}e^{j(\omega_e t + \theta_a)} = \frac{3}{2}NI_m e^{j(\omega_e t + \theta_a)} \tag{4.34}$$

This MMF space vector comes from the following expression:

$$\overrightarrow{F}(t) = \left[e^{j0}i_a(t) + e^{j\frac{2\pi}{3}}i_b(t) + e^{j\frac{4\pi}{3}}i_c(t)\right] \tag{4.35}$$

The general space vector of a three-phase variables $f_a(t), f_b(t), f_c(t)$ is defined as:

$$\overrightarrow{f(t)} = \frac{2}{3}\left[e^{j0}f_a(t) + e^{j\frac{2\pi}{3}}f_b(t) + e^{j\frac{4\pi}{3}}f_c(t)\right] \tag{4.36}$$

Note that the coefficient 2/3 is used.

If $f_a(t), f_b(t), f_c(t)$ are a balanced three-phase set with an amplitude as f_m, then the end result is

$$\begin{aligned}\overrightarrow{f(t)} &= \frac{2}{3}\left[e^{j0}f_a(t) + e^{j\frac{2\pi}{3}}f_b(t) + e^{j\frac{4\pi}{3}}f_c(t)\right] \\ &= f_m e^{j\theta_a} \qquad \text{This is the analytic form of } f_a\end{aligned} \tag{4.37}$$

In other words, the real part of the space vector is the signal of phase a.

Note that the analytic form of a signal is a complex-valued function that has no negative frequency components Gabor (1946). If $f_a(t) = f_m\cos(\theta_a)$, then its Hilbert transform can be defined as $f_a'(t) = f_m\sin(\theta_a)$. The analytic signal is

$$f_a(t) + jf_a'(T) = f_m(\cos(\theta_a) + j\sin(\theta_a) = f_m e^{j\theta_a}.$$

4.2.1 Example

When $t = t_1$,

$$\begin{cases} i_a(t_1) = 1 \\ i_b(t_1) = -0.5 \\ i_c(t_1) = -0.5 \end{cases} \tag{4.38}$$

find the air gap MMF due to the three-phase stator currents at the moment of t_1: $F(\alpha, t_1)$.

Solution: There are two approaches to solve this problem.

1) Substituting $\omega_e t + \theta_a = 0$ and $I_m = 1$ in (4.31) and (4.32), we have $F(\alpha, t_1) = \frac{3}{2}N\cos\alpha$.

2) Using the phasor diagram in Figure 4.4 to plot the answer. First, at a-axis, we plot $\overrightarrow{F}_a(t_1)$ in the same direction of a-axis and with a magnitude of N where N is the number of the windings. Next, at the opposite of the b-axis, we plot $\overrightarrow{F}_b(t_1)$ with a magnitude of $0.5N$. At the opposite of the c-axis, we plot $\overrightarrow{F}_c(t_1)$ with a magnitude of $0.5N$. The sum of the three vectors is $1.5N$ at the direction of a-axis. Therefore $F(\alpha, t_1) = 1.5N\cos\alpha$.

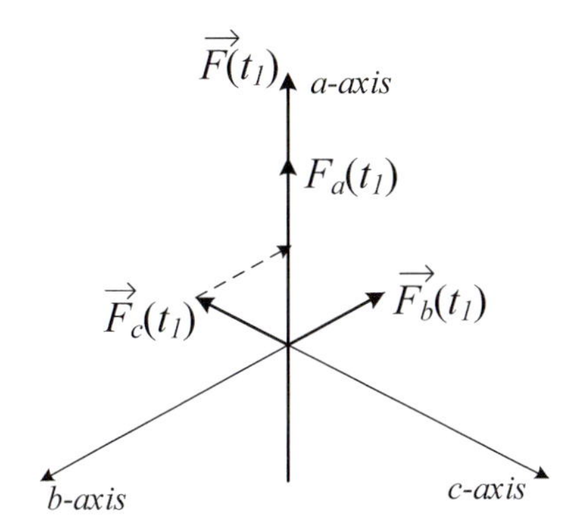

Figure 4.4: Space Vector diagram.

4.2.2 Advantages of space vector technique

Why is space vector so important? With space vectors, we can conduct decomposition and have a much better understanding regarding the analysis of machines with salient rotors. Like phasors, space vectors translate sinusoidal waveforms into vectors. Decomposition becomes very easy.

For example, we would like to decompose the MMF formed by the stator currents into two MMFs, one aligned with the d-axis and the other aligned with the q-axis.

We will first notate the MMF as $\overrightarrow{F_s} = \frac{3}{2}NI_m e^{j\theta_a}$, the position of the d-axis is $e^{j\theta}$, and the position of the q-axis is $e^{j(\theta-\frac{\pi}{2})}$.

Let's use the d-axis or q-axis as the reference instead. Then the MMF space vector should be written based on d-axis or q-axis respectively.

The MMF space vector is expressed as follows.

$$\overrightarrow{F_s} = \frac{3}{2}NI_m e^{j\theta_a} \quad \text{static reference} \tag{4.39}$$

This space vector should be $\overrightarrow{F_s}e^{-j\theta}$ if it is expressed based on the d-axis:

$$\overline{F}_{s1} = \frac{3}{2}NI_m e^{j(\theta_a-\theta)} = \underbrace{\frac{3}{2}NI_m \cos(\theta_a-\theta)}_{F_{sd}} + j\underbrace{\frac{3}{2}NI_m \sin(\theta_a-\theta)}_{-F_{sq}} \tag{4.40}$$

The MMF space vector should be $\overrightarrow{F_s}e^{-j(\theta-\pi/2)}$ if it is expressed based on the q-axis since the q-axis position relative to the static reference is $\theta-\pi/2$.

$$\overline{F}_{s2} = \frac{3}{2}NI_m e^{j(\theta_a-\theta+\frac{\pi}{2})} = \underbrace{-\frac{3}{2}NI_m \sin(\theta_a-\theta)}_{F_{sq}} + j\underbrace{\frac{3}{2}NI_m \cos(\theta_a-\theta)}_{F_{sd}} \tag{4.41}$$

We start to use $\overline{F}$ for notation. $\overline{F}$ is a static vector at steady-state since the electric frequency and the rotor speed are the same for synchronous generators and further at steady-state the current magnitude is constant. $\theta_a - \theta$ is a constant. $\overline{F}$ is termed as complex vector.

We shall have:

$$\overrightarrow{F_s} = \overline{F_{s1}} e^{j\theta} \tag{4.42}$$

$$= \overline{F_{s2}} e^{j(\theta - \frac{\pi}{2})} = \underbrace{-\frac{3}{2} N I_m \sin(\theta_a - \theta) e^{j(\theta - \frac{\pi}{2})}}_{\overrightarrow{F_{sq}}} + \underbrace{\frac{3}{2} N I_m \cos(\theta_a - \theta) e^{j\theta}}_{\overrightarrow{F_{sd}}} \tag{4.43}$$

Thus we have shown that a space vector can be easily decomposed into two space vectors orthogonal with each other.

$$\overrightarrow{F_s} = \overrightarrow{F_{ds}} + \overrightarrow{F_{qs}} \tag{4.44}$$

$$= (F_{sd} - jF_{sq}) e^{j\theta} \tag{4.45}$$

$$= (F_{sq} + jF_{sd}) e^{j(\theta - \frac{\pi}{2})} \tag{4.46}$$

4.2.3 Relationship of space vector, complex vector, $\alpha\beta$ and Park's transformation

The definition of a space vector can be written in the matrix/vector format as:

$$\overrightarrow{i} = \frac{2}{3} \begin{bmatrix} e^{j0} & e^{j\frac{2\pi}{3}} & e^{-j\frac{2\pi}{3}} \end{bmatrix} \begin{bmatrix} i_a \\ i_b \\ i_c \end{bmatrix} \tag{4.47}$$

We now consider two reference frames, the first $\alpha\beta$, and the second dq reference frame. The $\alpha\beta$-frame is a static frame with the β-axis leads the α-axis 90°: $\overrightarrow{i} = i_\alpha + j i_\beta$. Hence in the $\alpha\beta$ reference frame:

$$\begin{bmatrix} i_\alpha \\ i_\beta \end{bmatrix} = \frac{2}{3} \begin{bmatrix} 1 & \cos\frac{2\pi}{3} & \cos\frac{2\pi}{3} \\ 0 & \sin\frac{2\pi}{3} & -\sin\frac{2\pi}{3} \end{bmatrix} \begin{bmatrix} i_a \\ i_b \\ i_c \end{bmatrix} \tag{4.48}$$

In the dq reference frame, the reference axis is the d-axis, the space vector $\overrightarrow{i}$ in dq-frame becomes a new vector. We call this vector a complex vector and notate it as $\overline{I}_{dq} = i_d - j i_q$.

$$\overline{I}_{dq} = e^{-j\theta} \overrightarrow{i} = \frac{2}{3} \begin{bmatrix} e^{-j\theta} & e^{-j(\theta - \frac{2\pi}{3})} & e^{-j(\theta + \frac{2\pi}{3})} \end{bmatrix} \begin{bmatrix} i_a \\ i_b \\ i_c \end{bmatrix} \tag{4.49}$$

$$\begin{bmatrix} i_d \\ i_q \end{bmatrix} = \frac{2}{3} \begin{bmatrix} \cos\theta & \cos(\theta - \frac{2\pi}{3}) & \cos(\theta + \frac{2\pi}{3}) \\ \sin\theta & \sin(\theta - \frac{2\pi}{3}) & \sin(\theta + \frac{2\pi}{3}) \end{bmatrix} \begin{bmatrix} i_a \\ i_b \\ i_c \end{bmatrix} \tag{4.50}$$

For balanced three-phase currents with phase a current expressed as $I_m \cos(\theta_a)$, we can find i_d, i_q using the space vector concept. First the space vector for the current is $I_m e^{j\theta_a}$. We now view this space vector from the point of the rotor. The rotor's position is θ. Therefore, the complex vector in the dq frame is

$$\overline{I}_{dq} = I_m e^{j\theta_a} e^{-j\theta}.$$

This gives i_d and i_q as

$$\begin{aligned} i_d &= I_m \cos(\theta - \theta_a), \\ i_q &= I_m \sin(\theta - \theta_a). \end{aligned} \tag{4.51}$$

i_d is the current space vector's projection on the d-axis while i_q is the current space vector's projection on the q-axis.

To make the transformation matrix in (4.50) a square matrix, we add the zero sequence component where $i_0 = \frac{1}{3}(i_a + i_b + i_c)$. Then the $dq0$ variables have the following relationship with the abc variables.

$$\begin{bmatrix} i_d \\ i_q \\ i_0 \end{bmatrix} = \underbrace{\frac{2}{3} \begin{bmatrix} \cos\theta & \cos(\theta - \frac{2\pi}{3}) & \cos(\theta + \frac{2\pi}{3}) \\ \sin\theta & \sin(\theta - \frac{2\pi}{3}) & \sin(\theta + \frac{2\pi}{3}) \\ \frac{1}{2} & \frac{1}{2} & \frac{1}{2} \end{bmatrix}}_{T_1} \begin{bmatrix} i_a \\ i_b \\ i_c \end{bmatrix} \tag{4.52}$$

Textbooks on ac machines, e.g., Krause (1986), use this type of transformation. Bergen and Vittal (2009) uses a scaling factor $k = \sqrt{\frac{3}{2}}$ for the transformation matrix. Examine T_1,

$$T_1 T_1^T = \frac{4}{9} \begin{bmatrix} \frac{3}{2} & 0 & 0 \\ 0 & \frac{3}{2} & 0 \\ 0 & 0 & \frac{3}{2} \end{bmatrix} = \frac{2}{3}\mathbf{I} \tag{4.53}$$

Adding a scaling factor will make the transformation matrix an orthogonal or unitary matrix, that is

$$\underbrace{kT_1}_{T_2} kT_1^T = \mathbf{I} \tag{4.54}$$

Bergen and Vittal (2009) uses T_2 as the transformation matrix.

$$\begin{bmatrix} i'_d \\ i'_q \\ i'_0 \end{bmatrix} = \underbrace{\sqrt{\frac{2}{3}} \begin{bmatrix} \cos\theta & \cos(\theta - \frac{2\pi}{3}) & \cos(\theta + \frac{2\pi}{3}) \\ \sin\theta & \sin(\theta - \frac{2\pi}{3}) & \sin(\theta + \frac{2\pi}{3}) \\ \frac{1}{2} & \frac{1}{2} & \frac{1}{2} \end{bmatrix}}_{T_2} \begin{bmatrix} i_a \\ i_b \\ i_c \end{bmatrix} \quad (4.55)$$

Here we notate the variables based on T_2 transformations with $'$. The above transformation is called Park's transformation. In a nutshell, the stator related variables or space vectors are now viewed in the rotor's point of view after Park's transformation. In this text, complex vectors are adopted since they lead to simpler expressions.

$$\overline{I}_{dq} = e^{-j\theta}\overrightarrow{i} \quad (4.56)$$

$$\overline{v}_{dq} = e^{-j\theta}\overrightarrow{v} \quad (4.57)$$

$$\overline{\lambda}_{dq} = e^{-j\theta}\overrightarrow{\lambda} \quad (4.58)$$

Note only the current related space vector has a corresponding physical meaning related to MMF or flux. The rest has no physical meaning.

4.3 Synchronous generators with salient rotors

4.3.1 Armature reaction of a salient rotor generator

Compared to a round-rotor generator, the flux space vector formed in the air gap due to a salient-rotor generator's stator currents does not have a constant magnitude. This is due to the saliency of the rotor. At different rotor positions, the paths of flux lines will encounter different air gap distances. Chapter 7 of Bergen and Vittal (2009) gives an example to show the inductance of a static winding is a function of the rotor position. The three-phase circuits inductance matrix is also a function of rotor position θ. When Park's transformation is applied, the resulting dq-based flux linkages and currents are related with constant inductances.

With space vector decomposition technique, a simpler explanation can be offered.

Let's first decompose the MMF generated by the three-phase currents into two MMFs: one aligned with the d-axis ($\overrightarrow{F}_{sd}$) and the other aligned with the q-axis ($\overrightarrow{F}_{sq}$).

Based on the two MMFs, we can find the corresponding magnetic field density. If we examine the two paths of the flux lines, the air gap distances are fixed for these two MMFs regardless of rotor position: the air gap distance of d-axis flux lines is g_d, while the air gap distance for q-axis flux lines is g_q, and $g_d < g_q$. Therefore, the d-axis flux density should be stronger compared to the q-axis flux density if they are generated by the MMFs with the same magnitudes, or the reluctance of the q-axis magnetic circuit is higher than that of the d-axis. This eventually will lead to a greater flux linkage due to d-axis MMF, which means the related inductance L_d is greater than L_q, also $X_d > X_q$.

We should be able to find the expressions of $\overrightarrow{B_{sd}}$ and $\overrightarrow{B_{sq}}$ after a few steps. The first step is to find the amplitudes of the flux density, using Ampere's Law and considering only the air gap path. We have the following relationship.

$$\begin{aligned} 2g_d\frac{\hat{B}_{sd}}{\mu_0} &= F_{sd} = \frac{3}{2}NI_m\cos(\theta - \theta_a) = \frac{3}{2}Ni_d \\ 2g_q\frac{\hat{B}_{sq}}{\mu_0} &= F_{sq} = \frac{3}{2}NI_m\sin(\theta - \theta_a) = \frac{3}{2}Ni_q \end{aligned} \tag{4.59}$$

F_{sd} and F_{sq} can be viewed as the MMF space vector's projection on the d-axis and q-axis. Figure 4.5 presents the decomposition geometry.

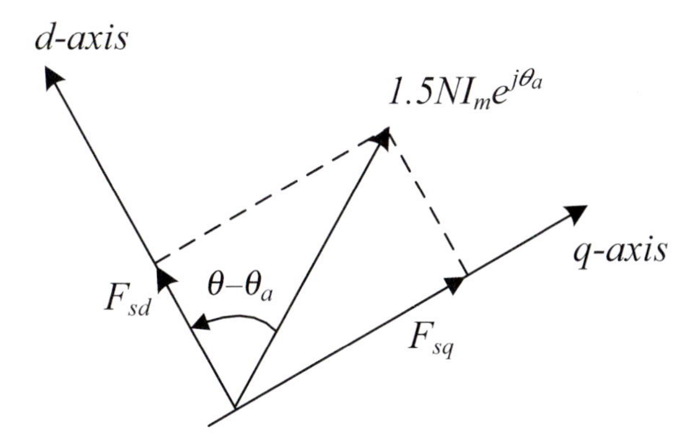

Figure 4.5: Stator MMF decomposition.

Then the space vectors of the flux density can be found.

$$\overrightarrow{B_{sd}} = \underbrace{\frac{4}{\pi}\frac{\mu_0}{2g_d}F_{sd}}_{B_{sd}}\, e^{j\theta} \tag{4.60}$$

$$\overrightarrow{B_{sq}} = \underbrace{\frac{4}{\pi}\frac{\mu_0}{2g_q}F_{sq}}_{B_{sq}}\, e^{j(\theta-\frac{\pi}{2})} \tag{4.61}$$

Consider the Gaussian space encompassed by the stator aa' winding (Figure 4.1). We will find the respective flux linkages:

$$\begin{aligned}
\lambda_{ad} &= 2NlrB_{sd}\cos(\theta) = 2Nlr\frac{4}{\pi}\frac{\mu_0}{2g_d}F_{sd}\cos(\theta) \\
&= \underbrace{2Nlr\frac{4}{\pi}\frac{\mu_0}{2g_d}\frac{3}{2}N}_{L_{d1}}\,\underbrace{i_d\cos(\theta)}_{i_{ad}} \\
\lambda_{aq} &= 2NlrB_{sq}\cos\left(\theta-\frac{\pi}{2}\right) = 2Nlr\frac{4}{\pi}\frac{\mu_0}{2g_q}F_{sq}\cos(\theta-\frac{\pi}{2}) \\
&= \underbrace{2Nlr\frac{4}{\pi}\frac{\mu_0}{2g_q}\frac{3}{2}N}_{L_{q1}}\,\underbrace{i_q\cos\left(\theta-\frac{\pi}{2}\right)}_{i_{aq}}
\end{aligned} \tag{4.62}$$

The total flux linkage linked to aa' is

$$\lambda_{aa'} = \lambda_{ad} + \lambda_{aq} = L_{d1}i_{ad} + L_{q1}i_{aq} \tag{4.63}$$

For round-rotor generators, $g_d = g_q$, therefore, $L_{d1} = L_{q1} = L_{s1}$. Hence

$$\lambda_{aa'} = \lambda_{ad} + \lambda_{aq} = L_{s1}(i_{ad} + i_{aq}) = L_{s1}i_a \tag{4.64}$$

Including the stator leakage, then

$$\begin{aligned}
\lambda_{ad} &= L_d i_{ad} \\
\lambda_{aq} &= L_q i_{aq}
\end{aligned} \tag{4.65}$$

L_d and L_q are defined by including the leakage inductance L_{ls}.

$$\begin{aligned}
L_d &= L_{d1} + L_{ls} \\
L_q &= L_{q1} + L_{ls}
\end{aligned} \tag{4.66}$$

4.3.2 Salient generator phasor diagram, power and torque

Voltage and current phasor diagram

For salient rotor generators, the open-circuit voltage is the same as that of a round-rotor generator. After we obtain the open-circuit voltage due to i_F and the flux linkages due to the armature reaction, we will now have the total flux linkage linked to stator phase a coil aa'.

$$\lambda_{ag} = \lambda_{aF} + \lambda_{ad} + \lambda_{aq} \tag{4.67}$$

where λ_{aF} is the flux linkage due to i_F and

$$\lambda_{aF} = \underbrace{2Nrl\frac{4}{\pi}\frac{\mu_0}{2g_d}N_F}_{M_F} i_F \cos(\theta) \tag{4.68}$$

If the number of winding turns on the rotor and the stator are the same ($N_F = N$), then we know that $L_{d1} = \frac{3}{2}M_F$.

$$\begin{aligned}
\lambda_{ag} &= \lambda_{aF} + \lambda_{ad} + \lambda_{aq} \\
&= M_F i_F \cos(\theta) + L_{d1} i_{ad} + L_{q1} i_{aq} \\
&= (M_F i_F + L_{d1} i_d)\cos(\theta) + L_{q1} i_q \cos\left(\theta - \frac{\pi}{2}\right)
\end{aligned} \tag{4.69}$$

Including the leakage, we now have

$$\lambda_a = (M_F i_F + L_d i_d)\cos\theta + L_q i_q \sin\theta \tag{4.70}$$

The space vector of the stator flux linkage can be expressed as:

$$\overrightarrow{\lambda} = \left[(M_F i_F + L_d i_d) - jL_q i_q\right] e^{j\theta} \tag{4.71}$$

Define the d-axis flux linkage magnitude as λ_d and q-axis flux linkage magnitude as λ_q. We have:

$$\overrightarrow{\lambda} = (\lambda_d - j\lambda_q)e^{j\theta}. \tag{4.72}$$

Then:

$$\begin{aligned}
\lambda_d &= M_F i_F + L_d i_d \\
\lambda_q &= L_q i_q
\end{aligned} \tag{4.73}$$

Based on the Faraday's Law, we can find the induced electromotive force (EMF) or the air gap voltage v_{ag}.

$$\begin{aligned} v_{ag} &= -\frac{d\lambda_{ag}}{dt} \\ &= \omega\left(M_F i_F + L_{d1} i_d\right)\sin\theta - \omega L_{q1} i_q \cos\theta \\ &= \omega\left(M_F i_F + L_{d1} i_d\right)\cos\left(\omega t + \theta_0 - \frac{\pi}{2}\right) - \omega L_{q1} i_q \cos(\omega t + \theta_0) \\ &= \sqrt{2}\left(\underbrace{\frac{\omega M_F i_F}{\sqrt{2}}}_{E_a} + \omega L_{d1}\underbrace{\frac{I_m}{\sqrt{2}}\cos(\theta - \theta_a)}_{I_{ad}}\right)\underbrace{\cos\left(\omega t + \theta_0 - \frac{\pi}{2}\right)}_{q-axis} \\ &\quad - \sqrt{2}\omega L_{q1}\underbrace{\frac{I_m}{\sqrt{2}}\sin(\theta - \theta_a)}_{I_{aq}}\underbrace{\cos(\omega t + \theta_0)}_{d-axis} \end{aligned} \tag{4.74}$$

where ω is the speed.

We will now start to use phasors to express (4.74). At steady-state, the speed is at nominal $\omega = \omega_0$.

$$\overline{V}_{ag} = E_a + X_{d1} I_{ad} - jX_{q1} I_{aq} \quad \text{based on } q\text{-axis} \tag{4.75}$$

Further, if we consider

$$\overline{I}_{ad} = jI_{ad}, \tag{4.76}$$

$$\overline{I}_{aq} = I_{aq} \tag{4.77}$$

based on q-axis, then (4.75) becomes

$$\begin{aligned} \overline{V}_{ag} &= \overline{E}_a - jX_{d1}(jI_{ad}) - jX_{q1} I_{aq} \\ &= \overline{E}_a - jX_{d1}\overline{I}_{ad} - jX_{q1}\overline{I}_{aq} \end{aligned} \tag{4.78}$$

Or

$$\overline{E}_a = \overline{V}_{ag} + jX_{d1}\overline{I}_{ad} + jX_{q1}\overline{I}_{aq} \tag{4.79}$$

Considering the stator resistance r and leakage reactance X_{ls}, then we will have the terminal voltage $\overline{V}_a$ and $\overline{V}_{ag}$ relationship.

$$\begin{aligned} \overline{V}_{ag} &= \overline{V}_a + (r + jX_{ls})\overline{I}_a \\ &= \overline{V}_a + r\overline{I}_a + jX_{ls}(\overline{I}_{ad} + \overline{I}_{aq}) \end{aligned} \tag{4.80}$$

Therefore,

$$\overline{E}_a = \overline{V}_a + r\overline{I}_a + j(X_{d1} + X_{ls})\overline{I}_{ad} + j(X_{q1} + X_{ls})\overline{I}_{aq} \tag{4.81a}$$
$$= \overline{V}_a + r\overline{I}_a + jX_q\overline{I}_{ad} + jX_q\overline{I}_{aq} + j(X_d - X_q)\overline{I}_{ad} \tag{4.81b}$$
$$= \overline{V}_a + r\overline{I}_a + jX_q\overline{I}_a + j(X_d - X_q)\overline{I}_{ad} \tag{4.81c}$$

We can make (4.81) be based on the q-axis; then the above equation becomes:

$$\begin{aligned} E_a &= \overline{V}_a e^{-j\delta} + r\overline{I}_a e^{-j\delta} + jX_d jI_{ad} + jX_q I_{aq} \\ &= \overline{V}_a e^{-j\delta} + r\overline{I}_a e^{-j\delta} + jX_d jI_{ad} - jX_q jI_{ad} + jX_q \underbrace{(jI_{ad} + I_{aq})}_{\overline{I}_a e^{-j\delta}} \\ &= (\overline{V}_a + (r + jX_q)\overline{I}_a)e^{-j\delta} - (X_d - X_q)I_{ad} \end{aligned} \tag{4.82}$$

The above relationship makes sure that

$$\overline{V'} = \overline{V}_a + (r + jX_q)\overline{I}_a = V' e^{j\delta} \tag{4.83}$$

Note that the phasor $\overline{V'}$ must be located at the q-axis since based on (4.82), $\overline{V'}e^{-j\delta}$ must have real value.

Therefore, given the terminal voltage and current phasors, we can use the above relationship to find the δ first. Then the current phasor will be decomposed into $\overline{I}_{ad}$ and $\overline{I}_{aq}$. Further E_a will be found.

The phasor diagrams are shown in Figure 4.6.

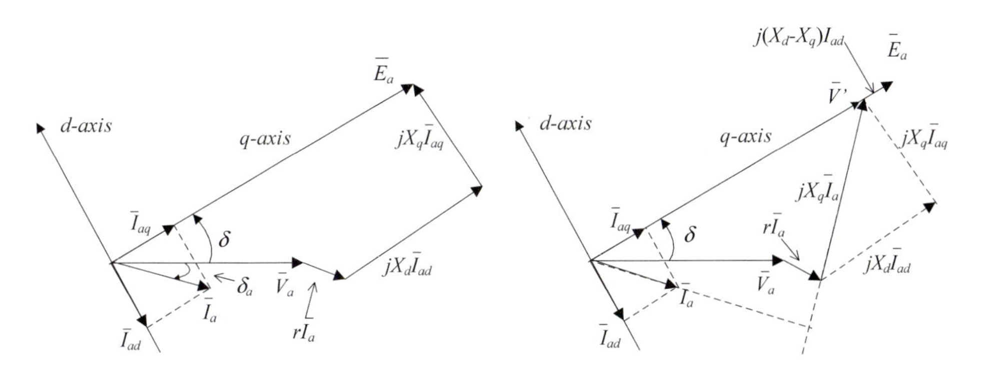

Figure 4.6: Phasor diagrams.

Further, for q-axis and d-axis respectively, we have the following relationship by separating the real and imaginary components of (4.81).

$$\begin{cases} E_a &= V_a \cos\delta + rI_a \cos(\delta_a - \delta) - X_d I_{ad} \\ 0 &= -V_a \sin\delta + rI_a \sin(\delta_a - \delta) + X_q I_{aq} \end{cases} \tag{4.84}$$

where $\theta = \omega t + \theta_0 = \omega t + \delta + \frac{\pi}{2}$, $\theta_a = \omega t + \delta_a$.

If we ignore r, we can find I_{ad} and I_{aq} more easily:

$$I_{aq} = \frac{V_a \sin\delta}{X_q} \tag{4.85}$$

$$I_{ad} = \frac{-E_a + V_a \cos\delta}{X_d} \tag{4.86}$$

Example: Short circuit analysis

Based on the phasor diagram, it can be easily found what is the short-circuit current when $V_a = 0$. In that case, the component contributed to E_a is completely dependent on $jX_d\overline{I}_{ad}$. Also since $V_a = 0$, the projection of $\overline{V}_a$ to the d-axis is zero. Therefore $V_a \sin\delta = 0$, which also means $jX_q\overline{I}_{aq} = 0$ or the q-axis current $I_{aq} = 0$.

$$\therefore \quad I_a = I_{ad} = \frac{E_a}{X_d} \tag{4.87}$$

Comparing the above short circuit analysis with that of the round rotor case, we find that it follows Lentz's Law as well. The induced flux should weaken the original flux. Therefore, the armature flux should be opposite to the rotor flux. Hence, the current should have only d-axis current. And the current is 90° lag $\overline{E}_a$. The phasor diagram is shown in Figure 4.7.

Power

Two points of view are given to explain the power expression.

Circuit point of view Given E_a, V_a, and the angle between them δ, we should be able to compute the complex power sent out from the generator.

$$S_a - \overline{V}_a \overline{I}_a^* \tag{4.88}$$

$$= V_a e^{-j\delta}(I_{aq} - jI_{ad}) \text{ if we use } \overline{E}_a\text{'s direction are the reference} \tag{4.89}$$

$$= V_a \left((I_{aq} + jI_{ad})e^{j\delta}\right)^* \text{ if we use } \overline{V}_a\text{'s direction are the reference} \tag{4.90}$$

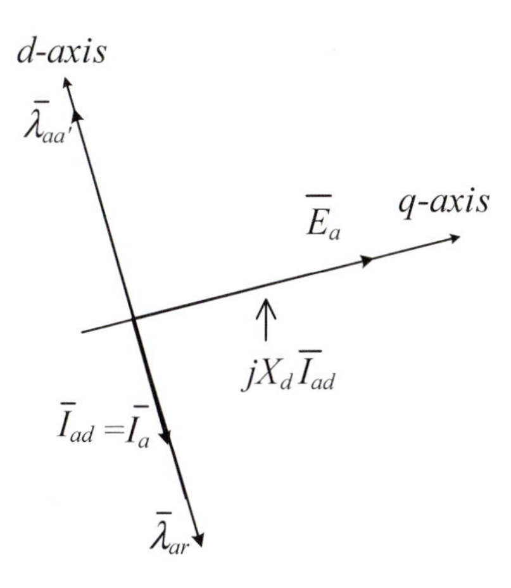

Figure 4.7: Phasor diagram when $V_a = 0$.

Either way, we will end up with

$$S_a = V_a(\cos\delta - j\sin\delta))(I_{aq} - jI_{ad}). \tag{4.91}$$

The real power from phase a should be

$$\begin{aligned} P_a &= V_a\cos\delta I_{aq} - V_a\sin\delta I_{ad} \\ &= \frac{E_a V_a}{X_d}\sin\delta + \left(\frac{1}{X_q} - \frac{1}{X_d}\right)\frac{V_a^2}{2}\sin(2\delta) \end{aligned} \tag{4.92}$$

The reactive power from phase a should be

$$\begin{aligned} Q_a &= -V_a\cos\delta I_{ad} + V_a\sin\delta I_{aq}) \\ &= \frac{E_a V_\infty}{X_d}\cos\delta - V_a^2\left(\frac{(\cos\delta)^2}{X_d} + \frac{(\sin\delta)^2}{X_q}\right) \end{aligned} \tag{4.93}$$

(4.92) shows that there are two components to generate torque or power. The first component is due to the rotor excitation. The second component is due to the saliency of the generator rotor. If the rotor is round, the second component is zero. This equation indicates that it is possible to generate torque or power without rotor excitation. In real-world applications, these types of machines are called reluctance machines.

Torque

Torque or power is produced on the windings with current flow inside a magnetic field. The basic relationship of force F, current I and magnetic field B is as follows:

$$\overrightarrow{F} = \overrightarrow{I} \times \overrightarrow{B} l \tag{4.94}$$

where l is the length of a winding. The direction of the force can be found using either the right-hand rule or the left-hand rule.

Right-hand rule: curl fingers from current direction to the flux direction. The force direction is the thumb's direction.

Left-hand rule: fingers point to the current direction while letting the palm face the flux lines. The thumb's direction is the force direction.

Figure 4.8 shows two examples to indicate the directions of the forces.

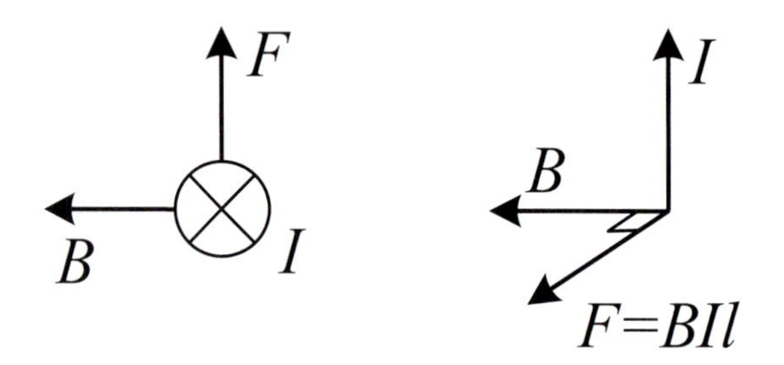

Figure 4.8: Force direction.

Next we show an example to find the torque in Figure 4.9. Note that the MMF direction generated by the current is α angle ahead of B. Note that the flux is decomposed into two elements, one aligned with the MMF (B_1), the other quadratic to the MMF (B_2). B_2 and I will not generate torque.

Assume that the default direction of torque is clockwise, i.e., when the MMF is leading the flux, torque will be positive. Otherwise, torque will be negative.

Then the torque computation is as follows.

$$T = F \cdot D = B_1 I l D = B I l D \sin\alpha \tag{4.95}$$

where α is angle of the MMF relative to the flux direction.

The above expression can also be obtained if we decompose the MMF or current into two components, one aligned with the flux line direction or the d-axis, notated as the I_d, and the other aligned with the q-axis, which lags the d-axis by 90°, notated as I_q. Then $I_d = I\cos\alpha$ and $I_q = -I\sin\alpha$ as shown in Figure 4.9. I_q will interact with B to generate torque while I_d will not interact with B to generate torque.

$$T = -B I_q l \cdot D = B I l D \sin\alpha. \tag{4.96}$$

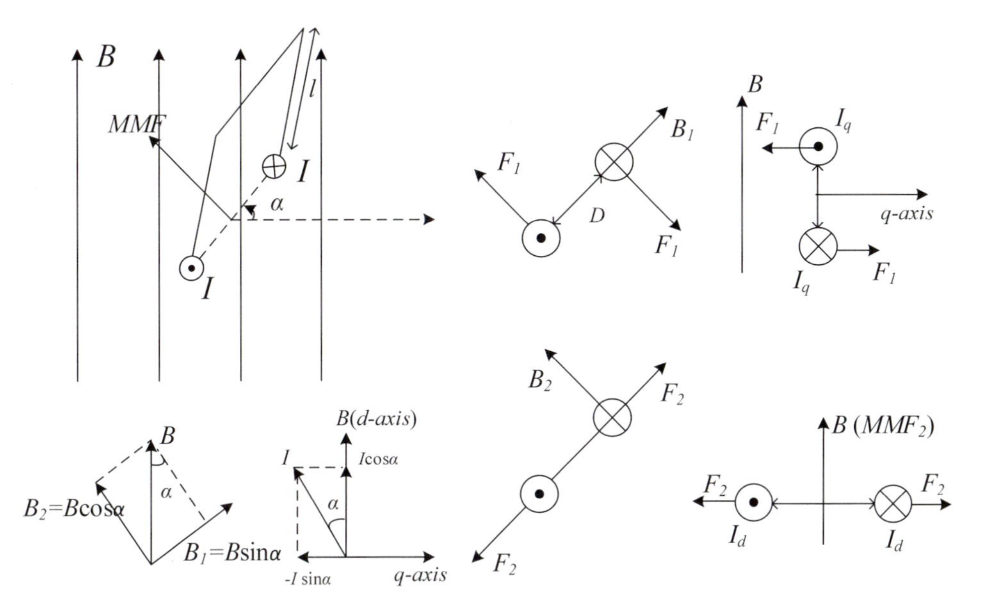

Figure 4.9: Torque generation. Note that the flux is decomposed into two elements, one aligned with the MMF (B_1), the other quadratic to the MMF (B_2). B_2 and I will not generate torque.

The negative sign is added since q-axis lags B by 90° and the generated torque is counterclockwise should I_q and B are all positive.

If we consider that there are N windings, then the torque should be

$$T = F \cdot D = NlDB_1I \tag{4.97}$$

Remarks: Torque is generated by the interactions of MMF and flux.

Round-rotor generator This principle is now extended to the rotating magnetic field. First we examine the round-rotor generator case. The rotor circuit will generate a magnetic field notated by $\overrightarrow{B}_F$ while the stator current forms a MMF $\overrightarrow{F}_s = \frac{3}{2}NI_m e^{j\theta_a}$.

When these two vectors are aligned with each other, no torque will be generated. This can also be corroborated by the fact that when $\overrightarrow{E}_a$ is 90 degree lagging or leading the current space vector, there is no power to be generated. To compute the torque, we can either decompose the rotor flux $\overrightarrow{B}_F$ into two components or decompose the stator MMF into two components, one aligned with the d-axis and the other aligned with the q-axis. Only the q-axis stator MMF will interact with the rotor flux to generate torque.

$$\begin{aligned}\vec{F}_{qs} &= F_s \sin(\theta - \theta_a) e^{j(\theta - \frac{\pi}{2})} \\ T_e &= 2rlB_F F_{qs} \\ &= 2rlNB_F \frac{3}{2} N \underbrace{I_m \sin(\theta - \theta_a)}_{i_q} \\ &= \frac{3}{2} M_F i_F i_q \end{aligned} \tag{4.98}$$

T_e is positive when the rotor flux leads the stator MMF. This is also a condition when a rotating ac machine is working as a generator.

The definitions of i_{ad} and i_{aq} are given in the previous section: i_{ad} is the component of the phase a current i_a that will generate a space vector aligned with d-axis, while i_{aq} is the component that will generate a space vector aligned with q-axis.

$$\begin{cases} i_a = i_{aq} + i_{ad} \\ i_b = i_{bq} + i_{bd} \\ i_c = i_{cq} + i_{cd} \end{cases} \tag{4.99}$$

Two space vectors will be generated:

$$\vec{i_q} = \frac{2}{3}\left(i_{aq} e^{j0} + i_{bq} e^{j\frac{2\pi}{3}} + i_{cq} e^{-j\frac{2\pi}{3}}\right) = i_q e^{j(\theta - \frac{\pi}{2})} \tag{4.100}$$

$$\vec{i_d} = \frac{2}{3}\left(i_{ad} e^{j0} + i_{bd} e^{j\frac{2\pi}{3}} + i_{cd} e^{-j\frac{2\pi}{3}}\right) = i_d e^{j\theta} \tag{4.101}$$

Taking the real parts of the above two space vectors, we have the following two components of the stator current in phase a:

$$\begin{aligned} i_{aq} &= i_q \sin\theta = I_m \sin(\theta - \theta_a) \sin\theta = \sqrt{2} I_{aq} \sin\theta, \\ i_{ad} &= i_d \cos\theta = I_m \cos(\theta - \theta_a) \cos\theta = \sqrt{2} I_{ad} \cos\theta. \end{aligned} \tag{4.102}$$

It is easy to see the relationship of the space vector amplitudes versus the RMS values of phasor components.

$$\begin{aligned} i_q &= \sqrt{2} I_{aq} = I_m \sin(\theta - \theta_a) \\ i_d &= \sqrt{2} I_{ad} = I_m \cos(\theta - \theta_a) \end{aligned} \tag{4.103}$$

From the above torque equation (4.98), we will find the power expression and further compare the expression with the one derived from the Thevenin circuit in Figure 4.3.

$$P_{3\phi} = \omega T_e = \frac{3}{2}\omega M_F i_F i_q = \frac{3}{2}\sqrt{2}E_a\sqrt{2}I_{aq} = 3E_a I_{aq} \tag{4.104}$$

Based on the phasor diagram for the salient rotor generator, the projection of the terminal voltage vector on d-axis is canceled by the q-axis current times X_q. For a round rotor, replace X_q by X_s. So we have

$$V_a \sin\delta = I_{aq}X_s \tag{4.105}$$

Therefore,

$$P_{3\phi} = 3\frac{E_a V_a}{X_s}. \tag{4.106}$$

This is the expression derived based on the Thevenin circuit when r is ignored.

Salient generator In the salient generator case, the q-axis stator MMF component will interact with the total flux due to rotor current and the d-axis stator MMF and generate torque while the d-axis stator MMF component will interact with the q-axis stator MMF. The two stator MMF components are shown as follows.

$$\begin{aligned}
\vec{F}_{qs} &= F_s\sin(\theta-\theta_a)e^{j(\theta-\frac{\pi}{2})} = \frac{3}{2}N\underbrace{I_m\sin(\theta-\theta_a)}_{i_q}e^{j(\theta-\frac{\pi}{2})} \\
\vec{F}_{ds} &= F_s\cos(\theta-\theta_a)e^{j\theta} = \frac{3}{2}N\underbrace{I_m\cos(\theta-\theta_a)}_{i_d}e^{j\theta}
\end{aligned} \tag{4.107}$$

Torque will be generated.

$$\begin{aligned}
T_e &= 2rlB_dF_{qs} - 2rlB_qF_{ds} \\
&= 2rl(B_F + B_{sd})\frac{3}{2}Ni_q - 2rlB_{sq}\frac{3}{2}Ni_d \\
&= 2rl\frac{4}{\pi}\frac{\mu_0}{2g_d}\left(N_Fi_F + \frac{3}{2}Ni_d\right)\frac{3}{2}Ni_q - 2rl\frac{4}{\pi}\frac{\mu_0}{2g_q}\frac{3}{2}Ni_q\frac{3}{2}Ni_d \\
&= \frac{3}{2}(M_Fi_F + L_di_d)i_q - \frac{3}{2}L_qi_qi_d
\end{aligned} \tag{4.108}$$

Since

$$\begin{aligned}
\lambda_d &= M_Fi_F + L_di_d, \\
\lambda_q &= L_qi_q,
\end{aligned}$$

we now have the torque expression as

$$T_e = \frac{3}{2}(\lambda_d i_q - \lambda_q i_d) \tag{4.109}$$

If the machine has $\frac{P}{2}$ pole pairs, then the torque expression becomes

$$T_e = \frac{3}{2}\frac{P}{2}(\lambda_d i_q - \lambda_q i_d). \tag{4.110}$$

In the above expressions, λ_d represents the total d-axis flux in the air gap linked to the stator circuits whose currents will generate an MMF in d-axis direction; λ_q represents the total q-axis flux in the air gap linked to the stator circuits whose currents generate a rotating MMF in q-axis direction with constant magnitude.

The stator circuits thus can be viewed as two fictitious rotor circuits: the d-axis circuit generates d-axis MMF $\overrightarrow{F_{ds}}$ and the DC current in the circuit is $\frac{3}{2}i_d$; the q-axis circuit generates q-axis MMF $\overrightarrow{F_{qs}}$ and the DC current in the circuit is $\frac{3}{2}i_q$.

The expression in (4.109) is used in electric machinery books often, e.g., Krause (1986). This expression is applicable to any rotating machine: synchronous or induction.

$$\begin{aligned} P_{3\phi} = \omega T_e &= \frac{3}{2}\underbrace{\omega M_F i_F}_{\sqrt{2}E_a}\underbrace{i_q}_{\sqrt{2}I_{aq}} + \frac{3}{2}(\omega L_d - \omega L_q)i_q i_d \\ &= 3E_a I_{aq} + 3(X_d - X_q)I_{ad}I_{aq} \end{aligned} \tag{4.111}$$

We can replace I_{aq} with $\frac{V_a \sin\delta}{X_q}$ and replace I_{ad} with $\frac{V_a \cos\delta - E_a}{X_d}$. The power expression becomes:

$$P_{3\phi} = 3\left(\frac{E_a V_a \sin\delta}{X_d} + \left(\frac{1}{X_q} - \frac{1}{X_d}\right)\frac{V_a^2}{2}\sin(2\delta)\right) \tag{4.112}$$

4.4 Generator model based on space vector

The main dynamic is Faraday's Law for electromagnetism. We have both rotor circuit and stator circuits to examine. For the rotor circuit, considering the resistance, the excitation voltage has the following relationship with the excitation current i_F and the flux linkage linked to the rotor circuit λ_F.

$$v_f = r_f i_F + \frac{d\lambda_F}{dt} \tag{4.113}$$

The expression of λ_F has not been discussed in the previous sections since we focus mainly on the stator circuits for steady-state analysis.

In the air gap, there are three fluxes: B_F due to the rotor, B_{sd} and B_{sq} due to the stator currents. For rotor circuit windings, the q-axis flux will not result in any flux linkages since the rotor current's MMF is 90 degrees from the q-axis flux. Only the d-axis flux will generate flux linkage on the rotor circuit. Therefore

$$\begin{aligned} \lambda_F &= N_F Dl(B_F + B_{sd}) = N_F Dl \frac{4}{\pi}\frac{\mu_0}{2g_d}\left(N_F i_F + N\frac{3}{2}i_d\right) \\ &= L_F i_F + M_F \frac{3}{2} i_d \end{aligned} \tag{4.114}$$

where D is the distance between the rotor wires. For round rotor $D = 2r$. If $N_F = N$ and ignoring the leakage, then $L_F = M_F$.

Next, we examine the stator circuits. The ultimate objective is to express the stator dynamics from the viewpoint of the rotor. Then both rotor and stator dynamics are viewed from the rotor's point. Since the space vector is the combination of the abc variables, the dynamics of space vectors are the same as the dynamics expressed in the abc-frame. Hence we have:

$$\vec{v} = -r\vec{i} - \frac{d\vec{\lambda}}{dt} \tag{4.115}$$

where $\vec{\lambda}$ is the space vector associated with the flux linkages associated with the stator's abc circuits. We will now express the above relationship by complex vectors.

$$\begin{aligned} \overline{V}_{dq}e^{j\theta} &= -r\overline{I}_{dq}e^{j\theta} - \frac{d\overline{\lambda}_{dq}e^{j\theta}}{dt} \\ &= -r\overline{I}_{dq}e^{j\theta} - j\dot{\theta}\overline{\lambda}_{dq}e^{j\theta} - \frac{d\overline{\lambda}_{dq}}{dt}e^{j\theta} \end{aligned} \tag{4.116}$$

Get rid of $e^{j\theta}$ at the both sides:

$$\overline{V}_{dq} = -r\overline{I}_{dq} - j\dot{\theta}\overline{\lambda}_{dq} - \frac{d\overline{\lambda}_{dq}}{dt} \tag{4.117}$$

Separate the real and imaginary components:

$$\begin{aligned} v_d &= -ri_d - \dot{\theta}\lambda_q - \frac{d\lambda_d}{dt} \\ v_q &= -ri_q + \dot{\theta}\lambda_d - \frac{d\lambda_q}{dt}. \end{aligned} \tag{4.118}$$

The task left to us is to define λ_d and λ_q adequately. From our previous analysis on stator flux linkage using space vector decomposition, we know that

$$\begin{aligned} \lambda_d &= M_F i_F + L_d i_d, \\ \lambda_q &= L_q i_q. \end{aligned} \tag{4.119}$$

The d-axis flux linkage is generated due to the rotor flux and d-axis armature flux, which is also related to d-axis stator current; the q-axis flux linkage is due to the q-axis armature flux only.

For a generator with only a rotor excitation circuit, the following is the complete dynamic model related to electromagnetic dynamics.

$$\text{rotor: } v_f = r_f i_F + \frac{d\lambda_F}{dt} \tag{4.120a}$$

$$\text{stator: } v_d = -r i_d - \dot{\theta}\lambda_q - \frac{d\lambda_d}{dt} \tag{4.120b}$$

$$v_q = -r i_q + \dot{\theta}\lambda_d - \frac{d\lambda_q}{dt} \tag{4.120c}$$

$$\text{where: } \lambda_F = L_F i_F + M_F \frac{3}{2} i_d \tag{4.120d}$$

$$\lambda_d = M_F i_F + L_d i_d \tag{4.120e}$$

$$\lambda_q = L_q i_q \tag{4.120f}$$

With the derived dynamic model, Bergen and Vittal (2009) presents two interesting applications: voltage buildup and short-circuit. The first application helps readers understand the time-constant T'_{d0} and the second application helps readers understand the transient reactance X'_d that is used in fault analysis. In the first example, we will adopt the space vector concept and relationship between a space vector and its corresponding time-domain signal to quickly find the time-domain expression of the terminal voltage. In the second example, we show how to find closed-form expressions using MATLAB symbolic toolbox.

4.4.1 Application 1: Voltage Buildup

A generator rotates at nominal speed ω_0 and the stator is open-circuited and the initial $i_F = 0$. The excitation voltage has a step response. Find the stator phase a voltage's time-domain expression.

Solution: The stator open circuit indicates that all stator currents are zero $i_d = i_q = 0$. Hence, the dynamic model in (4.120) in this case becomes the following.

$$\lambda_F = L_F i_F \tag{4.121a}$$
$$\lambda_d = M_F i_F \tag{4.121b}$$
$$\lambda_q = 0 \tag{4.121c}$$
$$\text{rotor: } v_f = r_F i_F + \frac{d\lambda_F}{dt} = r_F i_F + L_F \frac{di_F}{dt} \tag{4.121d}$$
$$\text{stator: } v_d = -\frac{d\lambda_d}{dt} = -M_F \frac{di_F}{dt} \tag{4.121e}$$
$$v_q = \dot{\theta}\lambda_d = \omega_0 M_F i_F \tag{4.121f}$$

From (4.121d), we can find out the time-domain expression of i_F, then we can find the expressions of v_d and v_q, further $v_a(t)$.

$$i_F = \frac{v_F}{r_F}(1 - e^{-(r_F/L_F)t})$$
$$v_d = -\frac{M_F v_F}{L_F} e^{-(r_F/L_F)t}$$
$$v_q = \frac{\omega_0 M_F v_F}{r_F}(1 - e^{-(r_F/L_F)t})$$

With the expression of v_d and v_q, we can find the complex vector and the space vector of the stator voltage:

$$\overline{V}_{dq} = v_d - jv_q = -\frac{M_F v_F}{L_F} e^{-(r_F/L_F)t} - j\frac{\omega_0 M_F v_F}{r_F}(1 - e^{-(r_F/L_F)t}) \tag{4.122}$$
$$\vec{v} = \overline{V}_{dq} e^{j\theta} = (v_d - jv_q)(\cos\theta + j\sin\theta) \tag{4.123}$$

The phase a voltage is the real part of the space vector:

$$v_a(t) = v_d \cos\theta + v_q \sin\theta \tag{4.124}$$

Compare v_d and v_q in (4.122), since $\frac{1}{L_F} \ll \frac{\omega_0}{r_F}$, v_d can be ignored and v_q is dominant.

$$v_a(t) \approx v_q \sin\theta = \frac{\omega_0 M_F v_F}{r_F}(1 - e^{-t/T'_{d0}}) \sin\theta \tag{4.125}$$

where $T'_{d0} = L_F/r_F$ is called the d-axis transient open-circuit time constant. The typical value is in the range of 2–9 seconds.
At steady-state,

$$v_a(t) \approx v_q \sin\theta = \frac{\omega_0 M_F v_F}{r_F} \sin\theta \tag{4.126}$$
$$= \sqrt{2} E_a \sin\theta \tag{4.127}$$

4.4.2 Application 2: Short-Circuit

The initial condition is open-circuit at the stator side. The rotor is rotating at the nominal speed and the excitation voltage v_F^0 is kept constant. At $t = 0^+$, the stator is connected to the ground. Ask for the time-domain expression of i_a.

Solution: For this case, a short-circuit indicates that the stator voltage is zero $v_d = v_q = 0$. Thus, the dynamic model is expressed as follows.

$$\text{rotor: } v_F = r_F i_F + \frac{d\lambda_F}{dt} \tag{4.128a}$$

$$\text{stator: } 0 = -r i_d - \dot{\theta}\lambda_q - \frac{d\lambda_d}{dt} \tag{4.128b}$$

$$0 = -r i_q + \dot{\theta}\lambda_d - \frac{d\lambda_q}{dt} \tag{4.128c}$$

$$\text{where: } \lambda_F = L_F i_F + M_F \frac{3}{2} i_d \tag{4.128d}$$

$$\lambda_d = M_F i_F + L_d i_d \tag{4.128e}$$

$$\lambda_q = L_q i_q \tag{4.128f}$$

Replacing the flux linkages by currents and conducting the Laplace transformation (the initial stator currents are zero), we have:

$$v_F = r_F i_F + L_F(s i_F - i_F^0) + \frac{3}{2} s M_F i_d \tag{4.129a}$$

$$0 = -r i_d - \omega_0 L_q i_q - M_F(s i_F - i_F^0) - s L_d i_d \tag{4.129b}$$

$$0 = -r i_q + \omega_0(M_F i_F + L_d i_d) - s L_q i_q \tag{4.129c}$$

In the matrix-vector format, we now have

$$\begin{bmatrix} r_F + sL_F & 1.5sM_F & 0 \\ -sM_F & -(r+sL_d) & -\omega_0 L_q \\ \omega_0 M_F & \omega_0 L_d & -(r+sL_q) \end{bmatrix} \begin{bmatrix} i_F \\ i_d \\ i_q \end{bmatrix} = \begin{bmatrix} v_F^0 + L_F i_F^0 \\ -M_F i_F^0 \\ 0 \end{bmatrix} = \begin{bmatrix} (r_F + L_F) i_F^0 \\ -M_F i_F^0 \\ 0 \end{bmatrix} \tag{4.130}$$

The current expressions can be found by Cramer's rule or using the MATLAB symbolic toolbox.

```
syms  s LF MF Ld Lq iF w0
A =[ s*LF, 1.5*s*MF, 0;
    -s*MF, -s*Ld, -w0*Lq;
    w0*MF, w0*Ld, -s*Lq];
b = [LF*iF; -MF*iF; 0];
```

```
i = inv(A)*b;
pretty(i)
```

Below is the answer given by the MATLAB symbolic toolbox assuming $r = r_F = 0$.

$$i_F(s) = \frac{(3M_F^2 - 2L_F L_d)s^2 - 2L_F L_d \omega_0^2}{\Delta} i_F^0 \tag{4.131}$$

$$i_d(s) = \frac{2L_F M_F \omega_0^2}{\Delta} i_F^0 \tag{4.132}$$

$$i_q(s) = \frac{\omega_0 M_F}{L_q(s^2 + \omega_0^2)} i_F^0 \tag{4.133}$$

where $\Delta = s(s^2 + \omega_0^2)(3M_F^2 - 2L_F L_d)$.

We now define $L_d' = L_d - \frac{3}{2}\frac{M_F^2}{L_F}$. Then

$$\begin{aligned}
\Delta &= s(s^2 + \omega_0^2) L_F L_d' \\
i_F(s) &= \left(\frac{L_d}{L_d'} \frac{1}{s} - \left(\frac{L_d}{L_d'} - 1 \right) \frac{s}{s^2 + \omega_0^2} \right) i_F^0 \\
i_d(s) &= -\frac{M_F i_F^0}{L_d'} \frac{\omega_0^2}{s(s^2 + \omega_0^2)} = -\frac{M_F i_F^0}{L_d'} \left(\frac{1}{s} - \frac{s}{s^2 + \omega_0^2} \right) \\
i_q(s) &= \frac{M_F i_F^0}{L_q} \frac{\omega_0}{s^2 + \omega_0^2}
\end{aligned}$$

The time-domain expressions are:

$$\begin{aligned}
i_F(t) &= \left(\frac{L_d}{L_d'} - \left(\frac{L_d}{L_d'} - 1 \right) \cos \omega_0 t \right) i_F^0 \\
i_d(t) &= -\frac{M_F}{L_d'} (1 - \cos \omega_0 t) i_F^0 = \frac{\sqrt{2} E_a}{X_d'} (\cos \omega_0 t - 1) \\
i_q(s) &= \frac{M_F i_F^0}{Lq} \sin \omega_0 t = \frac{\sqrt{2} E_a}{X_q} \sin \omega_0 t
\end{aligned}$$

The time-domain expression of $i_a(t)$ can be easily found based on $i_d(t)$ and $i_q(t)$.

$$\begin{aligned}
i_a(t) &= i_d \cos\theta + i_q \sin\theta \\
&= \frac{\sqrt{2} E_a}{X_d'} (\cos \omega_0 t - 1) \cos\left(\omega_0 t + \frac{\pi}{2} + \delta \right) + \frac{\sqrt{2} E_a}{X_q} \sin(\omega_0 t) \sin\left(\omega_0 t + \frac{\pi}{2} + \delta \right) \\
&= \frac{\sqrt{2} E_a}{X_d'} \sin(\omega_0 t + \delta) - \frac{E_a}{\sqrt{2}} \left[\frac{1}{X_d'} + \frac{1}{X_q} \right] \sin\delta - \frac{E_a}{\sqrt{2}} \left[\frac{1}{X_d'} - \frac{1}{X_q} \right] \sin(2\omega_0 t + \delta)
\end{aligned} \tag{4.134}$$

The above expression shows that $i_a(t)$ has three components: a DC component, a fundamental frequency component, and a second harmonic component. For the fundamental frequency component, the RMS value is E_a/X'_d.

4.5 Simplified dynamic model: flux decay model

Consider rotor dynamics only. The 3rd order electromagnetic dynamic model is now written as follows.

$$\text{rotor: } v_F = r_F i_F + \frac{d\lambda_F}{dt} \tag{4.135a}$$

$$\text{stator: } v_d = -ri_d - \dot{\theta}\lambda_q = -ri_d - \omega\lambda_q \tag{4.135b}$$

$$v_q = -ri_q + \dot{\theta}\lambda_d = -ri_q + \omega\lambda_d \tag{4.135c}$$

$$\text{where: } \lambda_F = L_F i_F + M_F \frac{3}{2} i_d \tag{4.135d}$$

$$\lambda_d = M_F i_F + L_d i_d \tag{4.135e}$$

$$\lambda_q = L_q i_q \tag{4.135f}$$

Note that the stator expressions are equivalent to the phasor diagram expression. $v_d = \sqrt{2}V_{ad}$ and $v_q = \sqrt{2}V_{aq}$, $i_d = \sqrt{2}I_{ad}$ and $i_q = \sqrt{2}I_{aq}$. The above relationship is the relationship between variables after Park's transformation and the phasors assume that $\omega = \omega_0$.

Using E_{fd} to replace v_F, for the rotor flux dynamics, we should have:

$$\text{rotor: } E_{fd} = \frac{\omega M_F}{\sqrt{2} r_F} v_F = \frac{\omega M_F}{\sqrt{2}} i_F + \frac{\omega M_F}{\sqrt{2}} \frac{L_F}{r_F} \frac{1}{L_F} \frac{d\lambda_F}{dt} \tag{4.136a}$$

$$\Rightarrow: \; E_{fd} = E_a + T'_{d0} \frac{dE'_a}{dt} \tag{4.136b}$$

$$\text{where: } \; E'_a = \frac{\omega M_F}{\sqrt{2}} \frac{\lambda_F}{L_F}, T'_{d0} = \frac{L_F}{r_F} \tag{4.136c}$$

E_{fd} and E'_a are introduced. E_{fd} can be viewed as the stator voltage corresponding to the rotor circuit voltage v_f while E'_a can be viewed as the stator voltage corresponding to the rotor flux λ_F. At steady-state, $E_{fd} = E_a$.

E_a, E'_a relationship

We now proceed to examine the relationship between E_a and E'_a. E_a is related to i_F while E'_a is related to λ_F. Therefore, the rotor flux linkage

λ_F's expression is listed as follows.

$$\lambda_F = L_F i_F + \frac{3}{2} M_F i_d \tag{4.137a}$$

$$\Rightarrow \frac{\omega M_F}{\sqrt{2} L_F} \lambda_F = \frac{\omega M_F}{\sqrt{2}} i_F + \frac{\omega M_F}{\sqrt{2} L_F} \frac{3}{2} M_F i_d \tag{4.137b}$$

$$\Rightarrow E'_a = E_a + \frac{3}{2} \frac{\omega M_F^2}{L_F} I_{ad} = E_a + (X_d - X'_d) I_{ad} \tag{4.137c}$$

$$\Rightarrow \overline{E'}_a = \overline{E}_a + \frac{\omega (k M_F)^2}{L_F} j(-j I_{ad}) \quad \text{aligned to the q-axis} \tag{4.137d}$$

$$= \overline{E}_a - j \frac{\omega (k M_F)^2}{L_F} \overline{I}_{ad} \tag{4.137e}$$

$$= \overline{E}_a - j(X_d - X'_d) \overline{I}_{ad} \tag{4.137f}$$

Note that X'_d is defined as $X'_d = X_d - \frac{\omega (k M_F)^2}{L_F}$, $k = \sqrt{\frac{3}{2}}$.

The above set of equations gives the relationship between E'_a and E_a. In addition, a phasor diagram is also given. Based on (4.137f), we have the following relationship of $\overline{E}'_a$, current $\overline{I}_a$ and the terminal voltage $\overline{V}_a$.

$$\overline{E}'_a = \overline{E}_a - j(X_d - X'_d) \overline{I}_{ad} \tag{4.138a}$$

$$= \overline{V}_a + r \overline{I}_a + j X_d \overline{I}_{ad} + j X_q \overline{I}_{aq} - j(X_d - X'_d) \overline{I}_{ad} \tag{4.138b}$$

$$= \overline{V}_a + r \overline{I}_a + j X'_d \overline{I}_{ad} + j X_q \overline{I}_{aq} \tag{4.138c}$$

The phasor diagram that shows the relationship among $\overline{E}_a$, $\overline{E}'_a$, $\overline{V}_a$ and $\overline{I}_a$ is presented in Figure 4.10.

The active power expression in terms of E'_a can also be found.

$$\begin{aligned} P_a &= V_a \cos\delta I_{aq} - V_a \sin\delta I_{ad} \\ &= \frac{E'_a V_a}{X'_d} \sin\delta + \left(\frac{1}{X_q} - \frac{1}{X'_d} \right) \frac{V_a^2}{2} \sin(2\delta) \end{aligned} \tag{4.139}$$

where δ is the angle between q-axis and the terminal voltage.

The flux decay model will be used to develop the plan model when the generator's automatic voltage regulator (AVR) is designed in Chapter 5.

Exercises

1. A round-rotor generator ($X_s = 1.0$, $r = 0.1$) is synchronized to a bus whose voltage is $1\angle 0°$. At synchronization $i_F = 1000$A (actual). The generator is then adjusted until $S_G = 0.8 + j0.6$. (S_G is the power supplied to

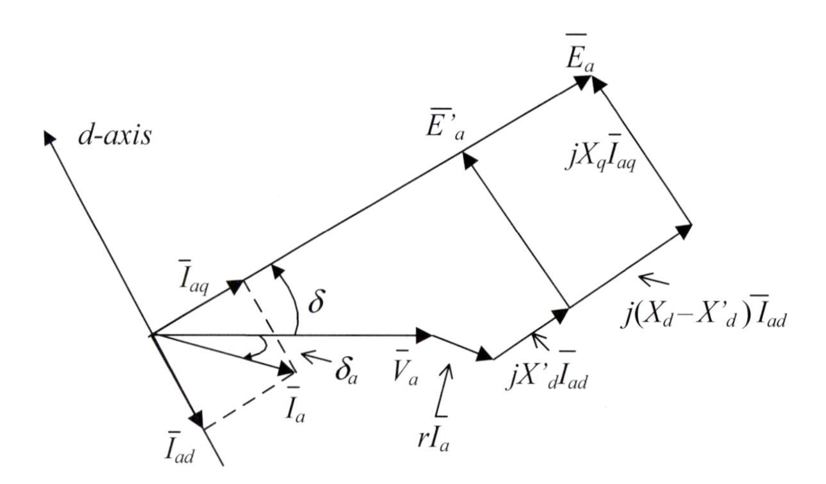

Figure 4.10: Phasor diagram.

the generator bus.).

1.1) Find i_F and the generator efficiency (assuming no generator loss except I^2R).

1.2) With the same i_F, what is the maximum active power the generator can deliver.

2. Consider a salient-pole generator delivering power through a short transmission line to an infinite bus. $\overline{V}_\infty = 1\angle 0°$, $E_a = 1.4$. The active power delivered to the infinite bus is 0.6. We are given the generator reactances $X_d = 1.6$ and $X_q = 1.0$ and the line reactance $X_L = 0.4$. Neglect resistances and find E_a and I_a.

3. For a set of balanced three-phase currents (with amplitude I_m, frequency ω, initial phase a angle 0), that is,

$$\begin{cases} i_a = I_m \cos(\omega t) \\ i_b = I_m \cos(\omega t - \frac{2\pi}{3}) \\ i_c = I_m \cos(\omega t + \frac{2\pi}{3}) \end{cases} \tag{4.140}$$

3.1) Find its space vectors at $t = 0$, $\omega t = \frac{2\pi}{3}$, and $\omega t = \frac{4\pi}{3}$.

If these are the stator currents for a generator, and assume the number of windings for each phase is N, the air gap permeability is μ_0, the length of the generator shaft is l, the radius from the center of the rotor to the air gap is r and the rotor is a round rotor (or the air gap has a uniform gap distance g, $g \ll r$).

3.2) Find the total MMF generated by the currents at a random place (α angle from the reference axis) in the air gap and a random time ($F(\alpha, t)$).

3.3) Find the magnetic flux density at angle α in the air gap $B(\alpha, t)$.

3.4) Find the total flux linkage linked to stator winding bb' ($\lambda_{bb'}(t)$), and its corresponding EMF $e_{bb'}(t)$.

4. For a salient rotor, the air gap distances are g_d and g_q for the d-axis and the q-axis. Decompose the above current space vector into d-axis and q-axis, assuming the rotor axis position is θ. Find

4.1) $\overrightarrow{i}$, $\overrightarrow{i_d}$ and $\overrightarrow{i_q}$.

4.2) MMF: $\overrightarrow{F_d}$, $\overrightarrow{F_q}$

4.3) Flux density: $\overrightarrow{B_d}$, $\overrightarrow{B_q}$

4.4) $\lambda_d(t)$ and $\lambda_q(t)$.

5. A synchronous generator in the steady-state is delivering power to an infinite bus. $\theta = \omega_0 t + \frac{\pi}{2} + \delta$, $\delta = \frac{\pi}{4}$, $\lambda_q = \lambda_d = \frac{1}{\sqrt{3}\omega_0}$, $i_q = \frac{1}{\sqrt{3}}$, $i_d = -\frac{1}{\sqrt{3}}$, $r = 0, X_d = \omega_0 L_d = 1, \omega_0 k M_F = 1, T'_{d0} = 1$ sec.
5.1) Find the torque (T_e. (b) Find $v_a(t), i_a(t)$, and i_F.
5.2) At $t = 0$, the generator is suddenly disconnected from the infinite bus. Assume that v_F =constant, $i_D = i_Q = 0$ (ignore damping circuits on the rotor). Sketch $i_F(t)$. Hint: use the fact that flux cannot jump to find $i_F(0^+)$.

6. Prove that a synchronous generator can be represented by a Thevenin equivalent: a voltage source E'_a behind an inductance $L'_d = L_d - \frac{(kM_F)^2}{L_F}$ if r_F is ignored. Remarks: This problem shows the magnetic field is viewed from

stator side as voltage source with E'_a as magnitude behind an inductance. When r_F is ignored, λ_F achieves steady-state with no time.

7. The shaft of a synchronous machine is clamped or "blocked" and is thus not free to turn. A set of positive-sequence voltage is applied to the motor terminals. Assume that (1)$\overline{V}_a = V_a e^{j0^\circ}$, (2) $\theta(t) = \frac{\pi}{2}$, (3)$i_F = 0$, (4)damping circuits not existing.

7.1) Find i_d and i_q in the steady-state.

7.2) Find an expression for the average torque $T_{e,av}$. Remarks: Torque contains a sinusoidal expression. Sinusoidal component does not generate energy. Therefore, only the DC component is sought.

7.3) Suppose that $r = 0$. What is $T_{e,av}$?

8. a SMIB system and assume that $X_d = 1.15, X_q = 0.6, X'_d = 0.15, X_L = 0.2, r = 0, T'_{d0} = 2$ sec. The generator is in steady-state with $E_{fd} = 1$and $E'_a = 1\angle 15^\circ$. Then at $t = 0$, E_{fd} is changed to a new constant value: $E_{fd} = 2$. Assume that the rotation is still uniform. Find $E'_a(t)$ for $t \geq 0$.

9. In MATLAB or Simulink, build a dynamic simulation model of a synchronous generator model with electromagnetic dynamics ONLY. This generator is connected to an infinitive bus. The initial condition is: $\bar{V}_\infty = 1\angle 0^o$, $\bar{I}_a = 1\angle 0^\circ$, Electromechanical dynamics can be ignored, i.e., you can assume the rotor speed is constant at $\omega_0 = 377$ rad/s and you don't need to put the swing dynamics into the model. You can opt to have just one-order dynamics (only rotor flux dynamics) or you can opt to have a third-order dynamic model by considering the dynamics of both rotor flux and qd stator flux linkages. You can even build a fifth-order dynamic model by considering the dynamics of rotor excitation circuit flux, D winding flux, Q winding flux, stator qd flux.

The parameters of the machine are given as follows:
kM_F can be found based on the relationship between L'_d, L_d, L_F and M_F.

9.1) Find the steady-state internal voltage $\bar{E}_a$, stator currents i_d, i_q, various flux linkages λ_d, λ_q, λ_F, as well as the rotor position θ against the stationary reference frame, and the position of q-axis relative to the phasor reference frame (rotating at ω_0 and at t=0, the phasor reference frame is aligned with the stationary reference frame).

X_d	1	L_d	$1/\omega_0$
X_q	0.8	L_q	$0.8/\omega_0$
X'_d	0.2	L'_d	$0.2/\omega_0$
X_{line}	0.5		
r_F	$0.5/\omega_0$	L_f	$1/\omega_0$

9.2) Show that the model has flat run for 1 second. Give a plot to show $E'_a(t)$ for 1 second or three subplots horizontally to show λ_d, λ_q and λ_F for 1 second.

9.3) Show the system dynamic response $E'_a(t)$ or flux linkages for a step change (10% increase) in E_{fd} or v_F. Show the dynamic response of the terminal voltage RMS magnitude $V_a(t)$. Plot the flat run (9.2) and the following dynamic responses in one plot.

Remarks: This exercise gives you an opportunity to learn dynamic model building. It includes two essential steps: initialization or flat run and dynamics due to differential equations. Initialization helps to calibrate the initial state variable values $x(0)$.

Chapter 5

Voltage Control

5.1 Introduction

The objective of voltage control is to have the root mean square (RMS) value of the terminal voltage of a generator (V_a) tracking a reference. As we have discussed from the electromagnetic model of a synchronous generator, the rotor excitation voltage v_F is treated as the input to the synchronous generator model. If we know the plant model $\frac{V_a}{v_F}$ (if the relationship is nonlinear, we need to obtain the linearized model $\frac{\Delta V_a}{\Delta v_F}$), we can then design the feedback control and test the controller accordingly. In the real world, design has to be carried out in multiple stages. Starting from the simplest plant model, we then add layers of complexity to see how those features will influence our design. In this particular case, we may need the following stages of model complexity for design.

1. Ignore all dynamics, only consider the steady-state relationship of v_F to V_a. At this stage, we examine two scenarios with the first as the simplest and the second with more complexity.

   - First scenario: stator side is open.
   - Second scenario: stator side is not open. We will consider a case of a single-machine infinite-bus (SMIB) system.

   This stage of plant model investigation helps to determine whether negative feedback or positive feedback will be employed.

2. Ignore electromechanical (EM) dynamics, only consider electromagnetic (EMT) dynamics. This includes multiple sub-stages.

- Consider only rotor flux dynamics. This model is termed "flux decay model."
- Consider rotor flux dynamics and stator flux dynamics for a generator with only an excitation circuit on the rotor. The three-order EMT model derived in Chapter 4 adopts the same assumption and can be used.
- Consider rotor flux dynamics (linked to the rotor excitation circuit), rotor damping circuit dynamics (linked to D and Q damping circuits), and stator flux dynamics. In this case, the dynamic model will have five orders. There will be five state variables, λ_F, λ_D, λ_Q, λ_d, λ_q. This model is a sophisticated model and is the base of the subtransient model.

3. Final stage: include both EM and EMT dynamics.

As we can see, the plant model can be very sophisticated. It is not possible and not necessary to include all dynamics. In the design stage, there is a compromise of plant model complexity and design simplicity. If a plant model becomes too complicated, insights may get lost. In addition, dynamics with very high bandwidth is not necessary to be included for control design at low bandwidth ranges. Therefore, a detailed comprehensive nonlinear model is only used in the validation or simulation stage.

In the generator voltage control design presented in Bergen and Vittal (2009), the modeling stops at the stage when the dynamics include both swing equation and rotor flux decay. In this book, the design also stops after including swing dynamics and rotor flux decay dynamics.

This chapter has four sections. Section 5.2 presents the simplest plant models represented by gains only. This investigation helps us to make a decision whether negative or positive feedback control should be adopted. Section 5.3 presents the plant models with rotor flux dynamics only. Electromechanical dynamics represented by swing equations are ignored. Section 5.4 examines voltage control design based on the plant models derived from Section 5.3. Finally Section 5.5 presents control design when both rotor flux dynamics and electromechanical dynamics are considered.

5.2 Plant model: No dynamics included

Through the phasor diagram or circuit analysis presented in Chapter 4, we should set up a concept that the terminal voltage V_a is influenced by the generator's internal voltage E_a, and E_a is proportional to the rotor excitation current i_F. The rotor excitation circuit voltage v_F can adjust i_F. Therefore, v_F can influence V_a. In this section, we seek the steady-state gain of $\frac{V_a}{v_F}$.

5.2.1 Scenario 1: Stator open-circuit

When the stator side is open, the stator currents are zero. Therefore, we can see that

$$V_a = E_a.$$

The steady-state value of the internal voltage E_a is related to the excitation current i_F. Further at steady-state, the excitation current can be found from the rotor circuit voltage v_F: $i_F = \frac{v_F}{r_F}$, where r_F is the resistance in the rotor circuit winding. Hence,

$$E_a = \frac{\omega M_F}{\sqrt{2}} i_F = \frac{\omega M_F}{\sqrt{2}} \frac{v_F}{r_F}. \tag{5.1}$$

Here, we would like to introduce a variable E_{fd}:

$$E_{fd} = \frac{\omega M_F}{\sqrt{2}} \frac{v_F}{r_F}. \tag{5.2}$$

E_{fd} will be used to replace v_F. E_{fd} can be viewed as the equivalent stator voltage due to the excitation voltage v_F.

According to (5.1), at steady-state, $E_{fd} = E_a$.

The plant model becomes

$$\frac{V_a}{v_F} = \frac{\omega M_F}{\sqrt{2} r_F}, \tag{5.3}$$

$$\text{or} \quad \frac{V_a}{E_{fd}} = 1. \tag{5.4}$$

5.2.2 Scenario 2: A SMIB system

In a more general case, stator currents are not zero. For the SMIB case, the grid voltage V_∞ is assumed to be constant. We will consider the grid voltage as the equivalent generator's terminal voltage while the equivalent generator's dq-axis reactances include the transmission line reactance.

Based on circuit analysis, we should be able to find dq-axis currents I_{ad} and I_{aq}, if E_a (or E_{fd}), V_∞ and δ (the angle δ between the q-axis of the generator and the reference synchronous rotating reference frame) are given. Further, based on the currents, we should be able to find V_a.

First, let us review the relationship of the dq-axis currents versus E_a and V_∞. A phasor diagram suited for the SMIB system is developed and shown in Figure 5.1. This phasor diagram is developed from the generator phasor diagram in Figure 4.10 presented in Chapter 4.

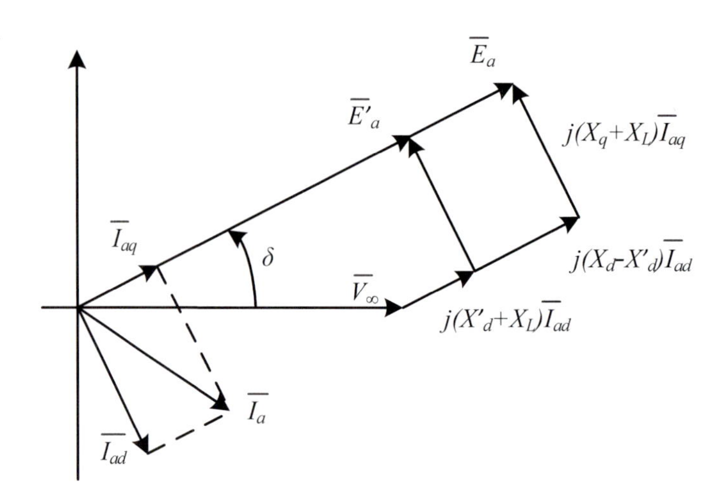

Figure 5.1: Phasor diagram.

$$\begin{aligned} I_{ad} &= \frac{V_\infty \cos\delta - E_a}{\widetilde{X}_d} \\ I_{aq} &= \frac{V_\infty \sin\delta}{\widetilde{X}_q} \end{aligned} \tag{5.5}$$

where $\widetilde{X}_d = X_d + X_L$, $\widetilde{X}_q = X_q + X_L$, and X_L is the reactance of the transmission line.

With that we should be able to find $\overline{V}_a$.

$$\overline{V}_a = V_\infty e^{j0} + jX_L(I_{aq} + jI_{ad})e^{j\delta} \tag{5.6}$$

We can decompose $\overline{V}_a$ into dq components.

$$\begin{aligned}
V_{aq} &= V_\infty \cos\delta + (jX_L)(jI_{ad}) = V_\infty \cos\delta - X_L I_{ad} \\
&= V_\infty \cos\delta - \frac{X_L}{\widetilde{X}_d}(V_\infty \cos\delta - E_a) \\
&= \frac{X_d}{\widetilde{X}_d} V_\infty \cos\delta + \frac{X_L}{\widetilde{X}_d} E_a \\
V_{ad} &= -V_\infty \sin\delta + X_L I_{aq} = -V_\infty \sin\delta + \frac{X_L}{\widetilde{X}_q} V_\infty \sin\delta \\
&= -\frac{X_q}{\widetilde{X}_q} V_\infty \sin\delta
\end{aligned} \tag{5.7}$$

V_a can be found based on

$$V_a = \sqrt{V_{aq}^2 + V_{ad}^2}. \tag{5.8}$$

The above analysis shows that V_a can be expressed by E_a using a non-linear function. For classical control design, linear plant models are desired. Therefore, small perturbation is applied to arrive at a linear model. The linear relationship of V_a with respect to E_a can be found as follows.

$$\begin{aligned}
\frac{\Delta V_a}{\Delta E_a} &= \frac{1}{\sqrt{V_{aq0}^2 + V_{ad0}^2}} \left(V_{aq0} \frac{\partial V_{aq}}{\partial E_a} + V_{ad0} \frac{\partial V_{ad}}{\partial E_a} \right) \\
&= \frac{V_{aq0} X_L}{V_{a0} \widetilde{X_d}}
\end{aligned} \tag{5.9}$$

V_{aq0} is the initial terminal voltage $\overline{V}_{a0}$'s projection on the q-axis. Normally, the angle between the q-axis and the terminal voltage space vector is less than 90°. Therefore, V_{aq0} is greater than zero.

Since at steady-state, $E_{fd} = E_a$, therefore, $\frac{\Delta V_a}{\Delta E_{fd}} = \frac{V_{aq0} X_L}{V_{a0} \widetilde{X_d}}$.

The above two scenarios show that the gain from E_{fd} to V_a is always positive. This plant model shows that a negative feedback control is justified.

In a negative feedback control system shown in Figure 5.2, error signals e will be generated by subtracting the measurements y from the reference values r. The plant G is a pure gain. k is also a pure gain.

If the terminal voltage is below its reference value, this error will be positive: $e > 0$. The error e will be passed to a controller, e.g., an amplifier. The output of the controller is the input of the plant u and $u > 0$ since $e > 0$. Since the plant is a pure gain, its output or measurements $y > 0$ if

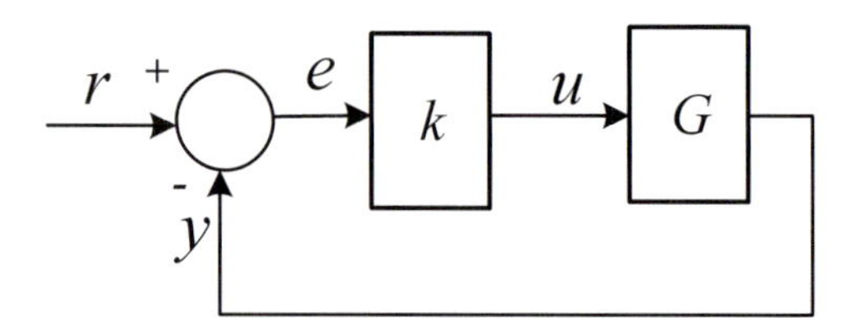

Figure 5.2: Negative feedback control: $e = \frac{1}{1+kG}r$ and $y = \frac{kG}{1+kG}r$.

$u > 0$. Therefore, an increasing positive error will result in an increase in the output of the plant y until $y = \frac{kG}{1+kG}r$. In the voltage control case, the terminal voltage will be increased if the measurement is reduced and the error is increased. Negative feedback control achieves the goal of increasing the terminal voltage in this case.

5.3 Plant model: Rotor flux dynamics only

The rotor flux dynamic is derived in Chapter 4.5 and is shown as follows.

$$E_{fd} = E_a + T'_{d0}\frac{dE'_a}{dt}, \tag{5.10}$$

where E_a is the internal voltage, E'_a is the stator voltage corresponding to the rotor flux linkage λ_F, E_{fd} is the stator voltage corresponding to the rotor voltage v_F, and T'_{d0} is notated as stator open-circuit time constant. 2 seconds is a typical value.

5.3.1 Stator open-circuit

For the stator open-circuit case, stator currents are zero. Therefore, $V_a = E_a$. In addition, the stator voltage corresponding to the rotor flux λ_F notated as E'_a is the same as the internal voltage E_a.

$$E'_a = \frac{\omega M_F}{\sqrt{2}}\frac{\lambda_F}{L_F} = \frac{\omega M_F}{\sqrt{2}} i_F = E_a,$$

since $\lambda_F = L_F i_F + \frac{3}{2}M_F i_d$ and $i_d = 0$. λ_F is due to i_F only.

Therefore, E_a can be replaced by E'_a in (5.10):

$$E_{fd} = E'_a + T'_{d0}\frac{dE'_a}{dt}. \tag{5.11}$$

The transfer function from E_{fd} to E'_a is as follows.

$$\frac{\Delta E'_a}{\Delta E_{fd}} = \frac{1}{T'_{d0}s+1} \tag{5.12}$$

Since $V_a = E_a = E'_a$, the transfer function from E_{fd} to V_a is

$$\frac{\Delta V_a}{\Delta E_{fd}} = \frac{1}{T'_{d0}s+1}. \tag{5.13}$$

5.3.2 SMIB case

In this case, the rotor flux dynamics equation includes E'_a and E_a. We need to express E_a as E'_a to have a first-order ODE. From that ODE, we will set up a transfer function from E_{fd} to E'_a. Then, based on the algebraic relationship between E'_a and V_a, we finally can arrive at the transfer function from E_{fd} to V_a.

Note in both steps, circuit analysis or phasor diagram analysis of the stator circuit help us find the algebraic relationships. Let us set out to analyze E'_a. As E'_a is related λ_F, and $\lambda_F = L_F i_F + \frac{3}{2}M_F i_d$, we can find the relationship between E'_a, E_a and the current i_d. In addition, for a SMIB case, $I_{ad} = \frac{V_\infty \cos\delta - E'_a}{\widetilde{X}'_d}$.

Based on the phasor diagram in Figure 5.1, we can find:

$$\begin{aligned} E_a &= E'_a + (\widetilde{X}_d - \widetilde{X}'_d)I_{ad} \\ &= E'_a + (\widetilde{X}_d - \widetilde{X}'_d)\frac{V_\infty \cos\delta - E'_a}{\widetilde{X}'_d} \\ &= \frac{\widetilde{X}_d}{\widetilde{X}'_d}E'_a + \left(\frac{\widetilde{X}_d}{\widetilde{X}'_d} - 1\right)V_\infty \cos\delta \end{aligned} \tag{5.14}$$

The small-perturbation model of the above relationship becomes:

$$\begin{aligned} \Delta E_a &= \frac{\widetilde{X}_d}{\widetilde{X}'_d}\Delta E'_a + \left(\frac{\widetilde{X}_d}{\widetilde{X}'_d} - 1\right)(-V_\infty \sin\delta)\Delta\delta \\ &= \frac{1}{K_3}\Delta E'_a - \left(\frac{1}{K_3} - 1\right)V_\infty \sin\delta\Delta\delta \\ &= \frac{1}{K_3}\Delta E'_a + K_4\Delta\delta \end{aligned} \tag{5.15}$$

where $K_3 \triangleq \frac{\widetilde{X}'_d}{\widetilde{X}_d}$ and $K_4 \triangleq 1 - \frac{1}{K_3}V_\infty \sin\delta$.

The rotor flux decay dynamics becomes

$$T'_{d0}\frac{d\Delta E'_a}{dt} = -\Delta E_a + \Delta E_{fd} = -\frac{1}{K_3}\Delta E'_a - K_4\Delta\delta + \Delta E_{fd} \tag{5.16}$$

$$\Longrightarrow (K_3T'_{d0}s+1)\Delta E'_a = K_3(-K_4\Delta\delta + \Delta E_{fd}) \tag{5.17}$$

If we ignore the electromechanical dynamics by assuming $\Delta\delta = 0$, then the transfer function from E_{fd} to E'_a is

$$\frac{\Delta E'_a}{\Delta E_{fd}} = \frac{K_3}{K_3T'_{d0}s+1} \tag{5.18}$$

With the electromechanical dynamics,

$$\Delta E'_a = \frac{K_3}{K_3T'_{d0}s+1}(\Delta E_{fd} - K_4\Delta\delta) \tag{5.19}$$

From E'_a to V_a

The terminal voltage $\overline{V}_a$ can be found from the phasor diagram.

$$\overline{E}'_a = \overline{V}_\infty + j\widetilde{X'_d}\overline{I}_{ad} + j\widetilde{X_q}\overline{I}_{aq} \tag{5.20}$$

Substituting $\overline{I}_{ad} = jI_{ad}e^{j\delta}$ and $\overline{I}_{aq} = I_{aq}e^{j\delta}$, we have the following equation based on the infinite bus reference frame.

$$E'_a e^{j\delta} = V_\infty + j\widetilde{X'_d}jI_{ad}e^{j\delta} + j\widetilde{X_q}I_{aq}e^{j\delta} \tag{5.21}$$

After multiplication by $e^{-j\delta}$, we have the following equation now with the q-axis as the reference.

$$E'_a = V_\infty\cos\delta - jV_\infty\sin\delta - \widetilde{X'_d}I_{ad} + j\widetilde{X_q}I_{aq}. \tag{5.22}$$

From the above equation, by equating real and imaginary parts, we have:

$$I_{ad} = \frac{V_\infty\cos\delta - E'_a}{\widetilde{X'_d}}, \tag{5.23}$$

$$I_{aq} = \frac{V_\infty\sin\delta}{\widetilde{X_q}}. \tag{5.24}$$

Let us now find V_a. The terminal voltage phasor can be expressed as follows.

$$\overline{V}_a = V_\infty + jX_L\overline{I}_a \tag{5.25}$$

Substituting $\overline{V}_a = (V_{aq} + jV_{ad})e^{j\delta}$ and $\overline{I}_a = (I_{aq} + jI_{ad})e^{j\delta}$, then multiplying by $e^{-j\delta}$ at both sides, we should have the following equation.

$$\begin{aligned} V_{aq} + jV_{ad} &= V_\infty \cos\delta - jV_\infty \sin\delta + jX_L(I_{aq} + jI_{ad}) \\ &= V_\infty \cos\delta - X_L I_{ad} + j(X_L I_{aq} - V_\infty \sin\delta) \end{aligned} \tag{5.26}$$

Now using (5.23) and (5.24) to substitute I_{ad} and I_{aq} in (5.26), we have:

$$V_{aq} = \frac{X_L}{\widetilde{X}'_d} E'_a + \left(1 - \frac{X_L}{\widetilde{X}'_d}\right) V_\infty \cos\delta \tag{5.27}$$

$$V_{ad} = \left(\frac{X_L}{\widetilde{X}_q} - 1\right) V_\infty \sin\delta \tag{5.28}$$

The RMS value can be found as

$$V_a = \sqrt{V_{aq}^2 + V_{ad}^2}. \tag{5.29}$$

V_a can be expressed by E'_a and δ in a nonlinear function.
Applying small perturbation, a linear expression is obtained as follows.

$$\Delta V_a = \frac{\partial V_a}{\partial \delta}\Delta\delta + \frac{\partial V_a}{\partial E'_a}\Delta E'_a \tag{5.30}$$

The partial derivative of V_a with respect to δ is defined as K_5 and the expression is as follows.

$$\begin{aligned} K_5 = \frac{\partial V_a}{\partial \delta} &= \frac{1}{\sqrt{V_{aq}^2 + V_{ad}^2}}\left(V_q \frac{\partial V_q}{\partial \delta} + V_d \frac{\partial V_d}{\partial \delta}\right) \\ &= \frac{V_\infty}{V_{a0}}\left[V_{aq0}\left(\frac{X_L}{\widetilde{X}'_d} - 1\right)\sin\delta_0 + V_{ad0}\left(\frac{X_L}{\widetilde{X}_q} - 1\right)\cos\delta_0\right] \\ &= -\frac{V_\infty}{V_{a0}}\left(\frac{V_{q0}X'_d}{\widetilde{X}'_d}\sin\delta_0 + \frac{V_{d0}X_q}{\widetilde{X}_q}\cos\delta_0\right) \end{aligned} \tag{5.31}$$

Similarly, we can find $K_6 = \frac{\partial V_a}{\partial E'_a}$ as follows.

$$\begin{aligned} K_6 = \frac{\partial V_a}{\partial E'_a} &= \frac{1}{\sqrt{V_{q0}^2 + V_{d0}^2}}\left(V_{q0}\frac{\partial V_q}{\partial E'_a} + V_{d0}\frac{\partial V_d}{\partial E'_a}\right) \\ &= \frac{V_{q0}X_L}{V_{a0}\widetilde{X}'_d} \end{aligned} \tag{5.32}$$

The small perturbation of V_a can be expressed as a linear function related to $\Delta\delta$ and $\Delta E'_a$.

$$\Delta V_a = K_5 \Delta\delta + K_6 \Delta E'_a \tag{5.33}$$

Ignoring $\Delta\delta$, the transfer function from E_{fd} to V_a can be written as

$$\frac{\Delta V_a}{\Delta E_{fd}} = \frac{\Delta E'_a}{\Delta E_{fd}} \frac{\Delta V_a}{\Delta E'_a} = \frac{K_3 K_6}{K_3 T'_{d0} s + 1} \tag{5.34}$$

The examination of the two scenarios shows that when the rotor flux dynamics are considered, the plant model becomes a first-order model. In the next section, we will show voltage control design based on the first-order plant models.

5.4 Voltage control design: Part I

In this section, the design is based on a generator with the stator side open. The electromechanical dynamics are ignored. This is same as the assumption that the generator is rotating at the nominal speed. The plant model is a first-order model: $\frac{1}{T'_{d0}s+1}$.

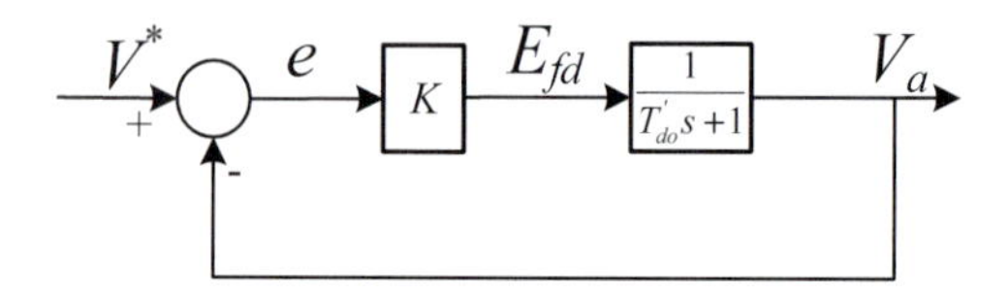

Figure 5.3: A simple feedback design of voltage controller.

The simplest design is shown in Figure 5.3 where the RMS of the terminal voltage is measured and compared with the reference voltage V^*. The error is then amplified by K times and E_{fd} is changed based on the output of the error. Instead of using integral control to bring the error to zero at steady-state, for voltage control, zero steady-state error is not a keen objective. Instead, a fast response is desired.

For the system in Figure 5.3, the closed-loop transfer function from V^* to the error e is expressed as:

$$\frac{e}{V^*} = \frac{1}{1 + \frac{K}{T'_{d0}s+1}} = \frac{T'_{d0}s + 1}{T'_{d0}s + K + 1}. \tag{5.35}$$

The closed-loop transfer function from V^* to the terminal voltage V_a is expressed as

$$\frac{V_a}{V^*} = \frac{\frac{K}{T'_{d0}s+1}}{1+\frac{K}{T'_{d0}s+1}} = \frac{K}{T'_{d0}s+K+1}. \tag{5.36}$$

Therefore, at steady-state, $V_a = \frac{K}{K+1}V^*$, $e = \frac{1}{K+1}V^*$. The bandwidth of the system is $\frac{K+1}{T'_{d0}}$. For $K = 1000, T'_{d0} = 1$, the bandwidth is 1000 rad/s, which indicates a 0.001 second time constant. Therefore, given a step response in V^*, it takes about 0.001 seconds for V_a to rise to 63% of its final steady-state value. A large gain K is desired to have a small error e as well as a fast response.

For this design model, we have not considered the dynamics of the excitation circuit, delay caused by actuator, and the amplifier dynamics. Considering those dynamics renders a more sophisticated model. In turn, those dynamics will pose limits on the gain.

5.4.1 Feedback control and the gain limit

Now we will consider the exciter, amplifier, and measurement unit's delay or dynamics. The block diagram is shown in Figure 5.4.

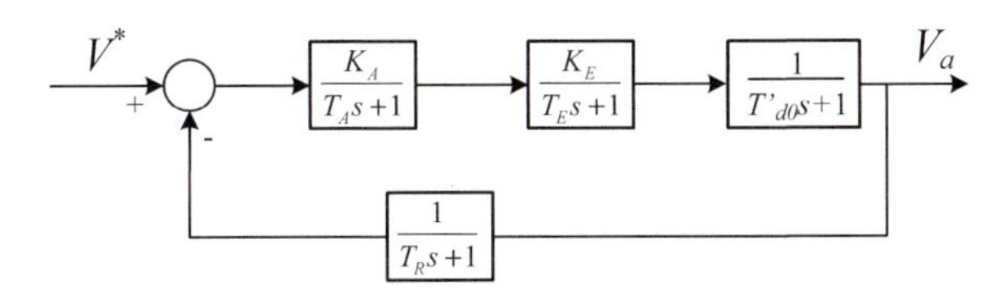

Figure 5.4: Voltage control block diagram considering dynamics of the exciter, amplifier and measurement unit.

The dynamics of an exciter can be expressed as $\frac{K_E}{T_E s+1}$. A typical value of $T_E = 0.8$ seconds and $K_E = 1$. The amplifier can be expressed as $\frac{K_A}{T_A s+1}$. The typical values are: $T_A = 0.05$ seconds. The measurement block is represented by $\frac{1}{T_R s+1}$. The typical value: $T_R = 0.06$ seconds.

The excitation block introduces an open-loop pole at $-\frac{1}{T_e} = -1.25$; the amplifier introduces an open-loop pole at $-\frac{1}{T_A} = -20$; and the measurement block introduces an open-loop pole close to $-\frac{1}{T_R} = -16.67$. Finally given $T'_{d0} = 5$, we have the fourth pole at -0.2 due to the generator.

Note that for this system, based on root locus analysis, there is a limit for the gain K_A. For this particular problem, K_A should be less than 57 as shown in the root loci in Figure 5.5.

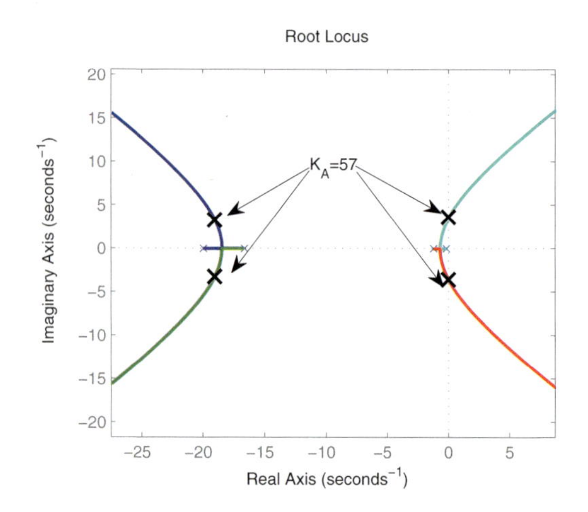

Figure 5.5: Root loci for the loop gain $\frac{1}{(1+T_As)(1+T_Es)(1+T'_{d0}s)(1+T_Rs)}$, where $T_A = 0.05$, $T_E = 0.8$, $T'_{d0} = 5$ and $T_R = 0.06$.

In order to have a small steady-state error between the voltage measurement and the voltage reference, we would like to have a large gain K_A. $K_A = 400$ is common in practice. The next topic is how to improve stability and increase the gain limit.

5.4.2 How to improve stability: Rate feedback

In order to have a small steady-state tracking error, a large K_A is preferred. In order to achieve stability with a large K_A, the control technique of rate feedback is employed. The rate feedback block is added and shown in Figure 5.6. The rate of E_{fd}: sE_{fd} will be used as a feedback signal to a proportional block with gain K_F. In control design, a pure derivative s is difficult to be realized exactly. The transfer function to realize derivative is $\frac{s}{T_Fs+1}$.

This part was discussed very briefly in Bergen and Vittal (2009). The book claims that "rate feedback" is a common practice in feedback control to enhance stability. In fact, we have seen rate feedback before. In frequency control, with and without droop control, the secondary frequency control gain limit are very different. With droop, then the gain can be large. Without droop, the gain has to be small. Compared to the integral frequency control, droop is a rate feedback. Later on, we will again show that a Power

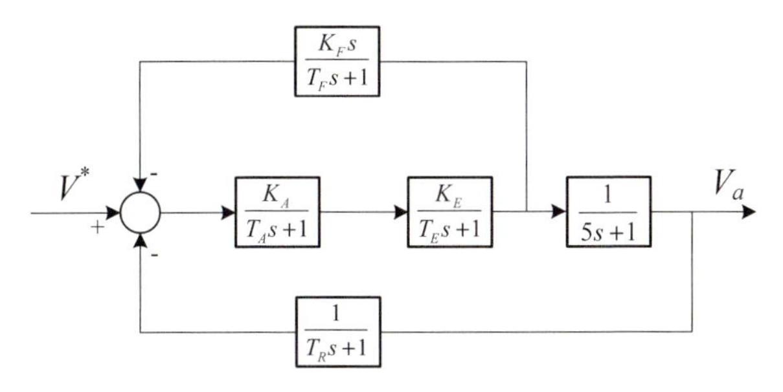

Figure 5.6: Block diagram including a rate feedback.

System Stabilizer (PSS) implements the concept of rate feedback to enhance damping.

What is the effect of rate feedback on root loci? Rate feedback introduces open-loop zeros to alter root locus paths. Zeros are used to attract poles that are close to the right half plane (RHP).

Shown in Figure 5.7, the open loop transfer function is re-examined for the system in Figure 5.6.

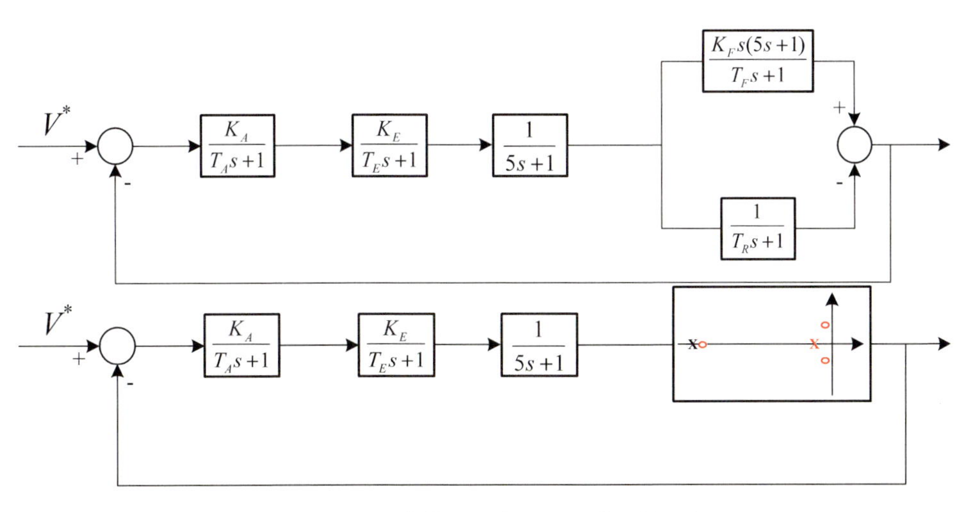

Figure 5.7: Effect of a rate feedback.

It can be seen that the effect of the rate feedback is to add a negative feedback of the derivative of E_{fd}. Equivalently, this is similar to adding three zeros (two complex conjugate, one at real-axis) and one pole on the loop gain of the open-loop system.

If $K_F = 0.1$, $T_F = 0.35$ and $T_R = 0.06$, the transfer function of $\frac{1}{T_R s+1} +$

$\frac{K_F s(5s+1)}{T_F s+1}$ becomes

$$\frac{0.03s^3 + 0.506s^2 + 0.45s + 1}{(1 + 0.35s)(1 + 0.06s))} = \frac{(s + 16.06)(s + 0.4023 \pm j1.38)}{(s + 2.86)(s + 16.67)} \tag{5.37}$$

Except that -16.67 is the pole related to the measurement unit, the rate feedback introduces three zeros and one pole as shown in Figure 5.7. The two zeros can attract the two poles that are close to RHP.

With rate feedback, the root loci for the loop gain are now shown in Figure 5.8. It can be seen that the gain of the amplifier K_A can be increased to 1000.

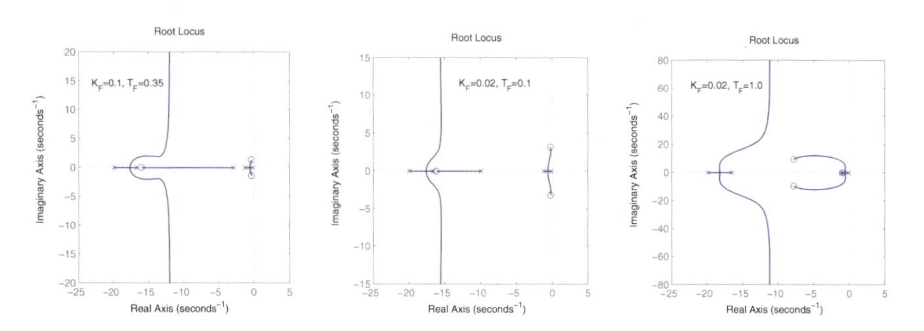

Figure 5.8: Root loci for the system in Figure 5.7 with rate feedback.

5.5 Voltage control design: Part II

5.5.1 Block diagram approach

In this section, the electromechanical dynamics are included. For this discussion, the SMIB case is considered as an example. For an open-circuit operating scenario, there is no power generated from the generator. With zero power, the rotating speed will be kept constant. Hence the electromagnetic and electromechanical are decoupled. However, when a generator sends power, the electromechanical dynamics is coupled with electromagnetic dynamics. As we can see from the previous section, ΔE_a, ΔV_a all have something to do with $\Delta\delta$.

In addition, the swing dynamics need to be re-examined. In frequency control discussed in Chapter 3, the generator model is assumed as a classical generator model with the internal voltage E_a constant. This is no longer the case when electromagnetic dynamics are considered. Here we have to rewrite the power expression in terms of $\Delta\delta$ and $\Delta E'_a$. In frequency control discussion, there is no $\Delta E'_a$ included (rotor flux dynamics not considered or electromagnetic not considered).

The power has been expressed by E'_a and δ in a nonlinear expression. Linearizing the expression will give us the linear model.

$$\Delta P_e = K_2 \Delta E'_a + T\Delta\delta \tag{5.38}$$

where $K_2 = \frac{\partial P_e}{\partial E'_a}$ and $T = \frac{\partial P_e}{\partial \delta}$.

The final system block diagram of a SMIB with automatic voltage regulator (AVR) is shown in Figure 5.9.

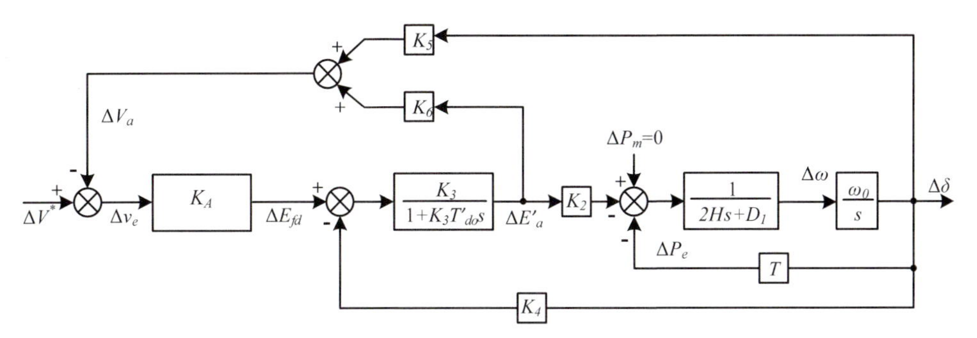

Figure 5.9: Block diagram of a SMIB system. The generator is equipped with AVR.

Note that the model of the voltage controller and the exciter are simplified to be a gain K_A only. The transfer function from ΔE_{fd} to ΔV_a can be found by examining the block diagram and through the usage of block manipulation. Interested readers can derive the transfer function manually.

5.5.2 State-space modeling approach

The transfer function can also be derived based on a state-space model. This approach is a systematic approach. First we will express the plant model as a state-space model $\dot{x} = Ax + Bu$ and $y = CX + Du$. Then we find the transfer function $Y(s)/U(s) = C(sI - A)^{-1}B + D$.

$$\dot{x} = \begin{bmatrix} \Delta\dot{\delta} \\ \Delta\dot{\omega} \\ \Delta\dot{E}'_a \end{bmatrix} = \begin{bmatrix} 0 & \omega_0 & 0 \\ \frac{-T}{2H} & \frac{-D_1}{2H} & \frac{-K_2}{2H} \\ \frac{-K_4}{T'_{d0}} & 0 & \frac{-1}{K_3 T'_{d0}} \end{bmatrix} \begin{bmatrix} \Delta\delta \\ \Delta\omega \\ \Delta E'_a \end{bmatrix} + \begin{bmatrix} 0 \\ 0 \\ \frac{1}{T'_{d0}} \end{bmatrix} \Delta E_{fd} \tag{5.39a}$$

$$y = \Delta V_a = K_5 \Delta\delta + K_6 \Delta E'_a = \begin{bmatrix} K_5 & 0 & K_6 \end{bmatrix} x \tag{5.39b}$$

The transfer function from ΔE_{fd} to ΔV_a is now found as follows.

$$\frac{\Delta V_a}{\Delta E_{fd}} = \begin{bmatrix} K_5 & 0 & K_6 \end{bmatrix} \begin{bmatrix} s & -\omega_0 & 0 \\ \frac{T}{2H} & s + \frac{D_1}{2H} & \frac{K_2}{2H} \\ \frac{K_4}{T'_{d0}} & 0 & s + \frac{1}{K_3 T'_{d0}} \end{bmatrix}^{-1} \begin{bmatrix} 0 \\ 0 \\ \frac{1}{T'_{d0}} \end{bmatrix} \tag{5.40a}$$

$$= \frac{K_3(2HK_6s^2 + D_1K_6s + \omega_0(K_6T - K_2K_5))}{2HK_3T'_{d0}s^3 + (2H + D_1K_3T'_{d0})s^2 + (D_1 + K_3TT'_{d0}\omega_0)s + (T + K_2K_3K_4)\omega_0} \tag{5.40b}$$

MATLAB codes to obtain the system model are shown as follows. The state-space model can be built using MATLAB function *ss* once A, B, C, D matrices are given as follows.

```
H = 3; D= 0;
T=1.01; K3=0.36; K4 = 1.47; K5 = -0.097; K6 = 0.417; K2 = 1;
w0 = 377; Td0 =2;
A =[0, w0, 0; -T/(2*H), -D/(2*H), -K2/(2*H); -K4/Td0, 0, -1/(K3*Td0)];
B =[0; 0; 1/Td0];
C = [K5, 0, K6];
G = ss(A,B, C, 0);
rlocus(G)
```

Root loci of the above system are shown in Figure 5.10. It can be seen from Figure 5.10 that if we design a feedback controller with V_a as the feedback signal, the gain K_A of the AVR cannot exceed 15.3. This example shows a more serious limitation on the gain of AVR: electromechanical dynamics. In real-world, power system stabilizers (PSSs) are used to improve the system stability. The limit of the gain can be relaxed after the installation of PSS.

5.5.3 Power system stabilizer

To improve the system stability, we examine again Figure 5.10. If we can move the two complex conjugate zeros to the left, the root loci can be kept at the left half plane (LHP). Therefore, the problem boils down to how to change the zero positions.

Since the transfer function is expressed as $C(sI - A)^{-1}B$ when D matrix is zero, zeros can be changed by adjusting C matrix, i.e., the output signals. In this case, let us try to change the output signal to be

$$\Delta V_a - k\Delta\omega = \begin{bmatrix} K_5 & -k & K_6 \end{bmatrix} \begin{bmatrix} \Delta\delta \\ \Delta\omega \\ \Delta E'_a \end{bmatrix}.$$

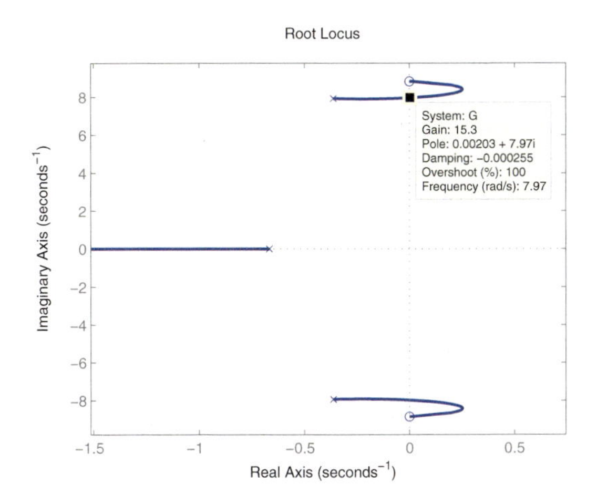

Figure 5.10: Root loci of the transfer functions in (5.40b).

The open-loop transfer function will have its zero positions changed. The additional signal added to V^* is $k\Delta\omega$. And this auxiliary control is named the power system stabilizer (PSS). The PSS output signal will be added to the reference voltage signal as shown in Figure 5.11.

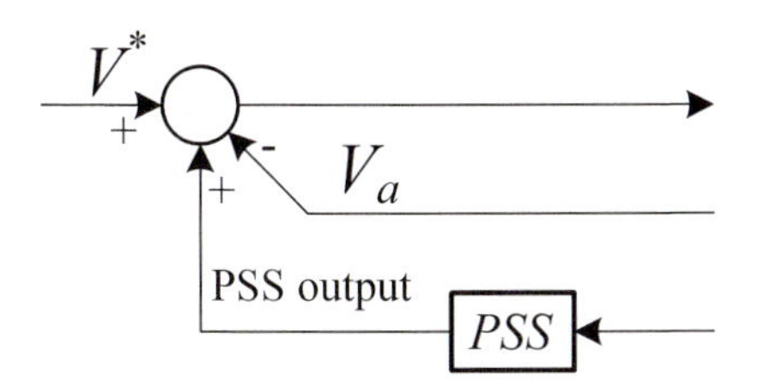

Figure 5.11: PSS implementation.

The transfer function is now changed as the following.

$$G_1 = \frac{K_3(2HK_6s^2+(DK_6+kK_2)s+\omega_0(K_6T-K_2K_5))}{2HK_3T'_{d0}s^3+(2H+DK_3T'_{d0})s^2+(D+K_3TT'_{d0}\omega_0)s+(T+K_2K_3K_4)\omega_0} \tag{5.41}$$

Note that the denominator does not change. However, the numerator has an additional item related to s. This change makes the open-loop system zeros move to the left. Letting $k = 10$, we find the root loci of G_1 in Figure 5.12. Figure 5.12 shows that the gain of the feedback voltage controller can now go infinity.

The above procedure is again explained by two approaches. In Approach

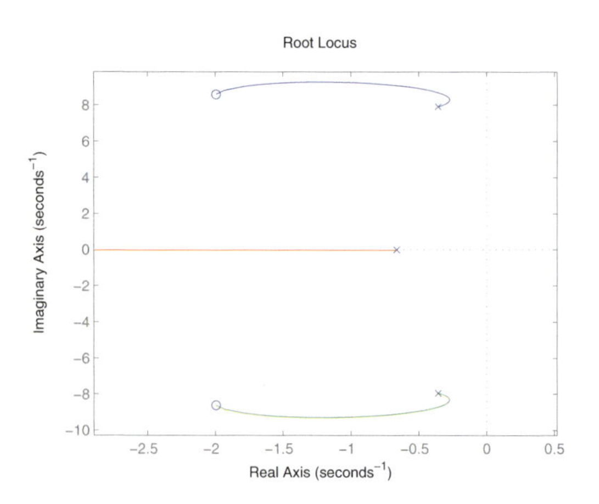

Figure 5.12: Root loci of the transfer functions in (5.41).

1, we examine how zeros can be changed. In Approach 2, we examine how to change the loop gain pole positions by adding a rate feedback.

Approach 1 Figure 5.13 presents the plant model from ΔE_{fd} to ΔV_a without PSS installed. Note that we can move the component of ΔV_a related to $\Delta E'_a$ ($K_6 \Delta E'_a$) forward to start at $\Delta\delta$. This is acceptable as long as we assume $\Delta P_m = 0$. Therefore, $K_6 \Delta E'_a$ is equivalent to $-\frac{K_6}{K_2}(Ms^2 + Ds + T)\Delta\delta$. We then combine one feedback loop into a transfer function $\frac{\Delta\delta}{\Delta E_{fd}} = -G_{F\delta}$. The pole zero maps for the two series connected blocks are presented in Figure 5.13.

Next, we consider moving zeros to LHP to enhance stability. This is done by adding the rate of the rotor angle in a forward block as shown in Figure 5.14. This forward block is PSS.

So the output y is now a combined signal with ΔV_a and the output of the PSS. The zeros of the system have been pushed to the left if we have the added component as $-ks\Delta\delta$. The negative sign is due to the negative coefficient of the highest order of $\frac{\Delta V_a}{\Delta\delta}$: $-\frac{K_6}{K_2}M$. In order to move the zeros to the left, $-ks$ is used to have the damping D increased. The new damping is now $D + \frac{kK_2}{K_6}$.

Approach 2 In the second approach, we again use the concept of rate feedback to first move the open-loop poles to left. This is done by adding a rate feedback at $\Delta\delta$ shown in Figure 5.15. The rate feedback is added to the input reference signal instead of subtracted from the reference signal. This is

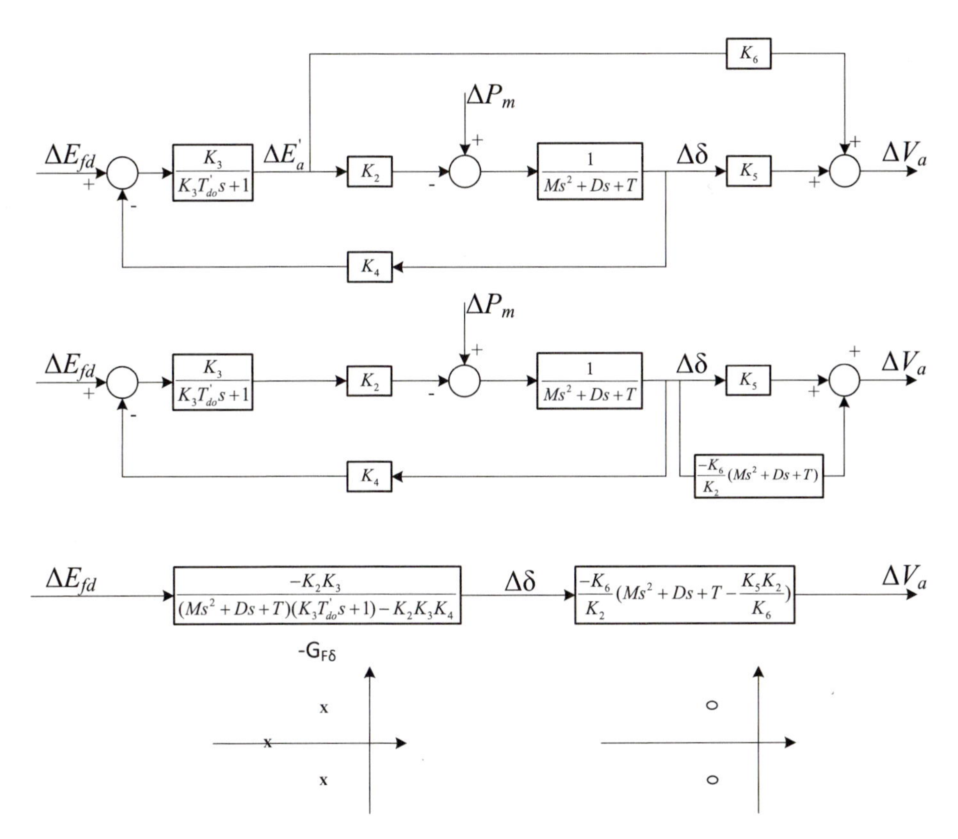

Figure 5.13: Plant model considering electromechanical dynamics.

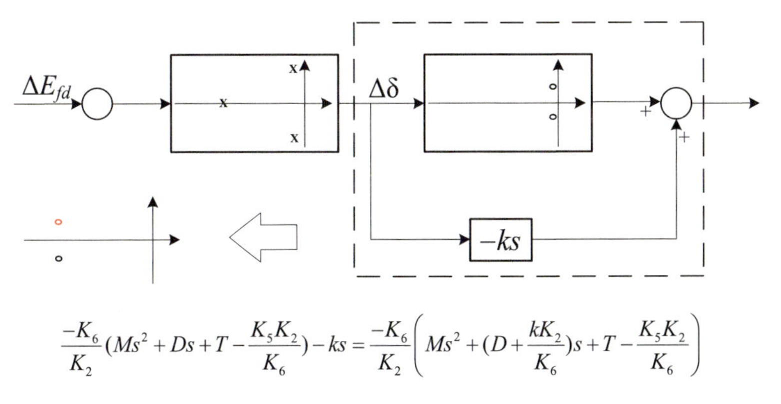

$$\frac{-K_6}{K_2}(Ms^2+Ds+T-\frac{K_5K_2}{K_6})-ks=\frac{-K_6}{K_2}\left(Ms^2+(D+\frac{kK_2}{K_6})s+T-\frac{K_5K_2}{K_6}\right)$$

Figure 5.14: Move zero to LHP.

due to the plant transfer function $\frac{\Delta\delta}{\Delta E_{fd}}=-G_{F\delta}$ has a negative steady-state gain $\frac{-K_2K_3}{T-K_2K_3K_4}$.

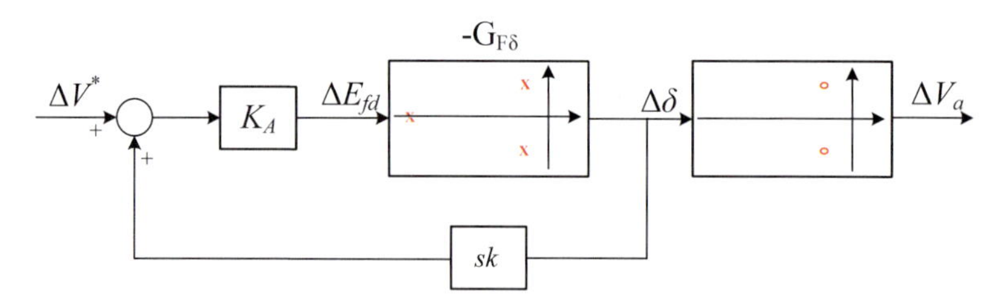

Figure 5.15: Rate feedback as PSS.

The effect of the rate feedback is illustrated in Figure 5.16. Note that the open-loop poles are moved to the left. The root loci will no longer pass the imaginary axis, which indicates enhanced stability.

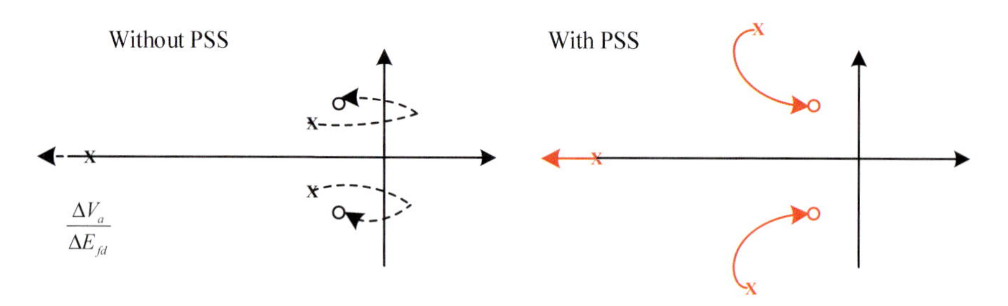

Figure 5.16: Effect of PSS.

Figure 5.17 shows the positions of the open-loop poles with and without rate feedback. A numerical case study is used to plot the root loci for the open-loop system $sG_{F\delta}$. The resulting root loci indicate the closed-loop system pole position. This closed-loop system is $\frac{\Delta\delta}{\Delta V^*}$ with rate feedback considered shown in Figure 5.15. $G_{F\delta}$ itself has only three poles. The open-loop zero shown in Figure 5.17 is introduced due to the rate feedback. It can be seen that for the closed-loop system, the two complex conjugate closed-loop system poles are moved to left. The new positions are notated as $*$ when $K_A = 400$.

A numerical case study on root loci for the loop gain from ΔE_{fd} to ΔV_a with rate feedback or PSS integrated are shown in Figure 5.18. This figure is a numerical example of the root loci illustration presented in Figure 5.16. This figure shows that with PSS ($k = 10/\omega_0$), K_A can be very large and the system is still stable.

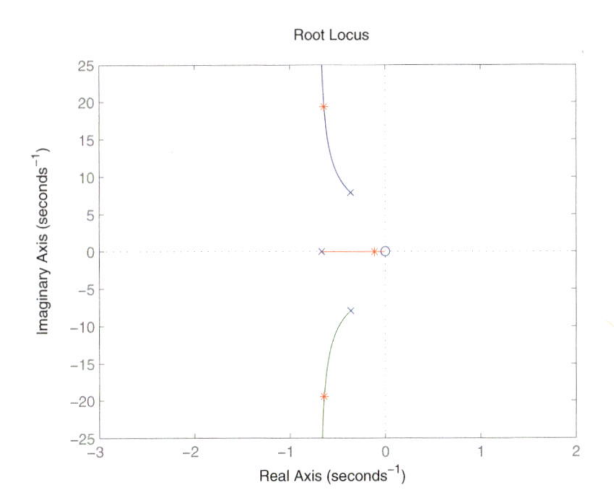

Figure 5.17: With rate feedback, closed-loop poles moved to the left for the system $\frac{\Delta\delta}{\Delta V^*}$. The loop gain used to plot root loci is $sG_{F\delta}$.

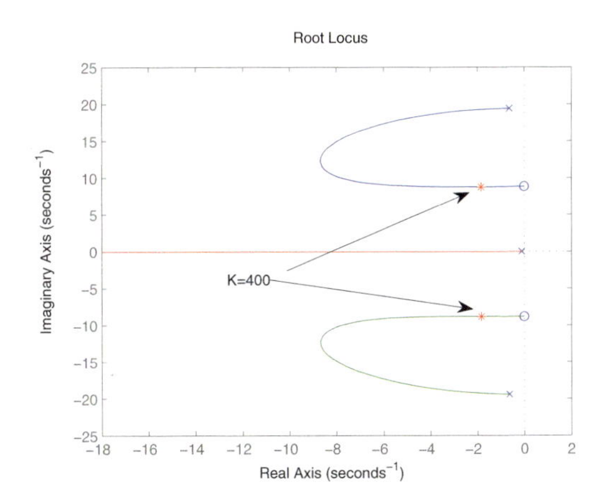

Figure 5.18: Root loci for $\frac{\Delta V_a}{\Delta E_{fd}}$ in Figure 5.14.

5.5.4 Linear model simulation results

In this subsection, time-domain simulation will be conducted in MATLAB/Simulink. The small-signal models will be built in Simulink. Figure 5.19 presents the MATLAB/Simulink blocks for two systems, one without PSS and one with PSS. The subsystem block shown in Figure 5.19 is presented in Figure 5.20.

Figure 5.21 presents the comparison of V_a when the rate feedback in the voltage control is ignored. For a system without PSS, it is unstable.

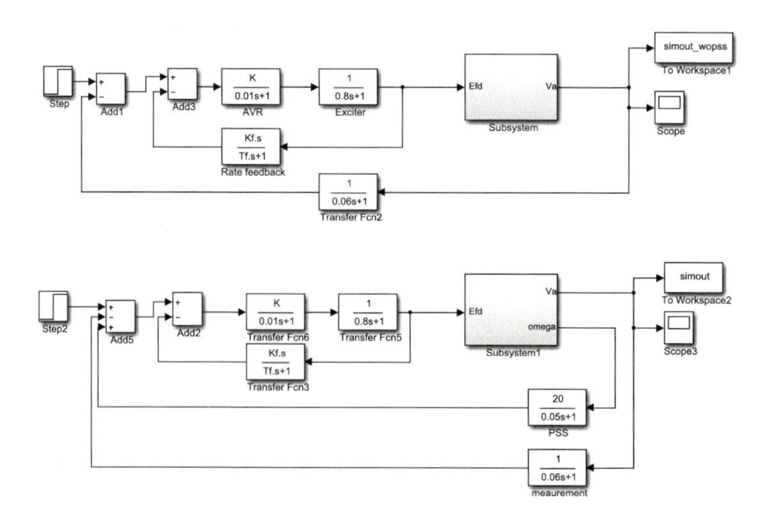

Figure 5.19: Simulink blocks. The first system has no PSS. The second system has PSS installed and the input signal to the PSS is the speed. The transfer function for the PSS is $\frac{20}{0.05s+1}$.

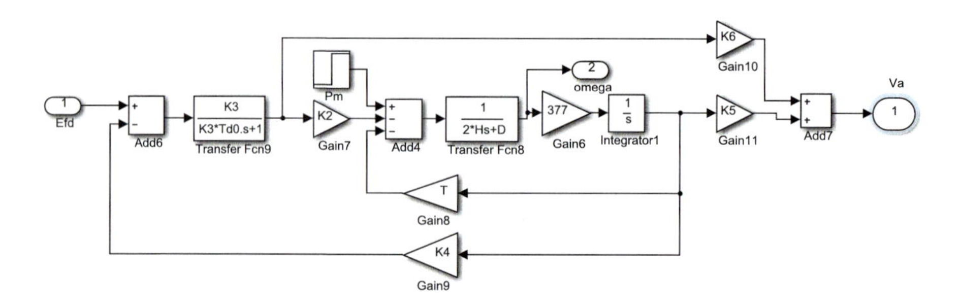

Figure 5.20: Simulink blocks for the system $\frac{\Delta V_a}{\Delta E_{fd}}$.

Finally, we examine the systems with voltage control rate feedback added. For the nominal system, the two show little difference. However, when power transfer increases T is reduced to 1/3 of its original value, the effect of PSS on stability is shown in Figure 5.22.

5.6 Summary

This chapter discusses generators' automatic voltage control and PSS design. The control design presented in this chapter demonstrates how to choose

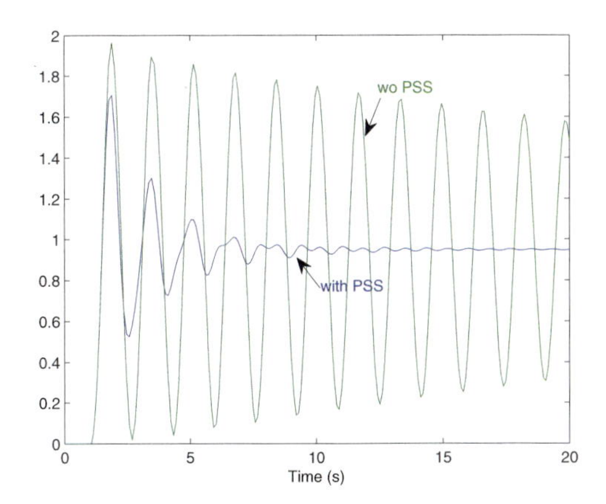

Figure 5.21: Step responses of ΔV_a for a voltage reference change. Both systems have no exciter voltage rate feedback.

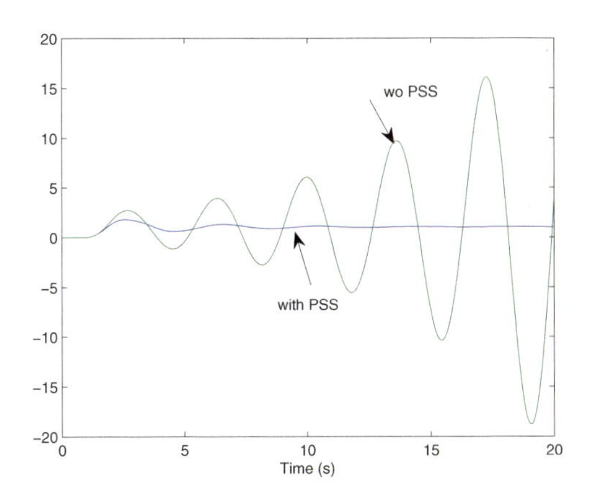

Figure 5.22: Step responses of ΔV_a for a voltage reference change for systems with exciter voltage rate feedback with heavy power transfer level.

negative feedback or positive feedback, when to use pure gain feedback, and how to improve stability using rate feedback. The application is voltage control through a generator's excitation voltage. The main analysis tool used in this chapter is the root locos method while the main validation tool is time-domain simulation.

Exercises

1 For a SMIB system, the transmission line reactance $X_L = 0.5$ pu and the generator parameters are as follows. $X_d = 1.0$, $X_q = 0.6$. The grid voltage is 1 pu and the generator sends out unity power factor power. $P_e = 0.5$ pu. Using these parameters, write codes to compute K_1 to K_6 for a given δ. Examine the power transfer level's impact (varying δ from -60 degrees to +60 degrees with 5 degrees as the step size: $-\frac{\pi}{3} : \frac{\pi}{18} : \frac{\pi}{3}$.) on K_i.

- Give a figure with five subplots to show a varying delta's influence on K_i (except K_3). K_1 is the same as T.
- Remark the effect of K_5 on stability. Show that heavy power transfer leads to negative K_5 and further stability issue by giving two root locus plots (one for $K_5 > 0$ and the other for $K_5 < 0$).

2. In MATLAB/Simulink, build a linearized model as shown in Figure 5.19.

- Compare the system dynamic performance due to a step response from ΔV_{ref} (use 10% change) with and without PSS.
- Use the MATLAB *rlocus* function to validate Figure 5.18. Please write the transfer function of the loop gain to which *rlocus* will be applied.
- Use the MATLAB function *linmod* to obtain the state-space model and the system matrix A. Eigenvalues of A are the poles of the closed-loop system. You can also find the closed-loop transfer function by manual derivation. Identify the closed-loop eigenvalues for the system with and without PSS. Compare your eigenvalues and linear system simulation results and state if they corroborate with each other.

Chapter 6

Frequency and Voltage Control in Microgrids

In this chapter, frequency and voltage control through voltage source converters (VSCs) in a microgrid will be discussed. VSC is the key element in a microgrid to interface distributed energy sources. The fundamental control structure of a VSC will be presented first. Then coordination among converters for real power and reactive power sharing will be presented. An alternative name for coordination is primary frequency control and primary voltage control.

In this chapter, we will examine two parallel VSCs, in either grid-connected mode or autonomous mode.

VSC's control and modulation are separately designed. The most popular modulation scheme is called pulse width modulation (PWM). Inputs to the PWM are the reference voltages while the outputs of the PWM are the switching sequences for VSC's gates. Though the three-phase voltages are discrete signals due to the implementation of the switching sequences, FFT analysis will show that the discrete waveform consists of the fundamental waveform (same as the reference voltage) and high-frequency components at the range of the switching frequency (about 1kHz). With a small L filter, the high frequency components can be filtered out. Therefore, the output voltage of a VSC can be viewed as sinusoidal. The details regarding switching and PWM can be found in Power Electronic books Mohan and Undeland (2007).

In this chapter, a VSC is viewed as a controllable voltage source with its *abc* three-phase voltages controllable. The VSC controls will adjust the three-phase voltage.

6.1 Control of a voltage source converter (VSC)

Cascaded control structure is popularly used for VSC control. The inner loop realizes current control and the outer loop realizes power control or voltage/frequency control. The reason to have an inner current control is that VSCs are sensitive to large currents. With current control, VSCs can be protected against overcurrents. The inner loop control and the outer loop control are usually designed separately due to their distinct bandwidth requirements. The inner current controls require a fast response and a high bandwidth, e.g., 200 rad/s. The outer controls require a much slower response and a lower bandwidth, e.g., 5 rad/s.

In the following paragraph, the *dq*-frame based VSC control structure will be explained. Interested readers can refer to Yazdani and Iravani (2010) for a thorough coverage on a single VSC's control, including both *dq*-frame based control and $\alpha\beta$-frame based control.

6.1.1 Design of inner current controller

The inner current controls for the vector control should be designed to be much faster than the outer control loops. The converter voltage in *abc* frame is notated as v and the current is notated as i. The voltage at the point of the common coupling (PCC) is notated as v_1. An RL circuit is considered between the converter and the PCC as shown in Figure 6.1.

Therefore:

$$L\frac{d\vec{i}}{dt} + R\vec{i} = \vec{v} - \vec{v}_1. \tag{6.1}$$

where $\vec{\cdot}$ is the space vector.

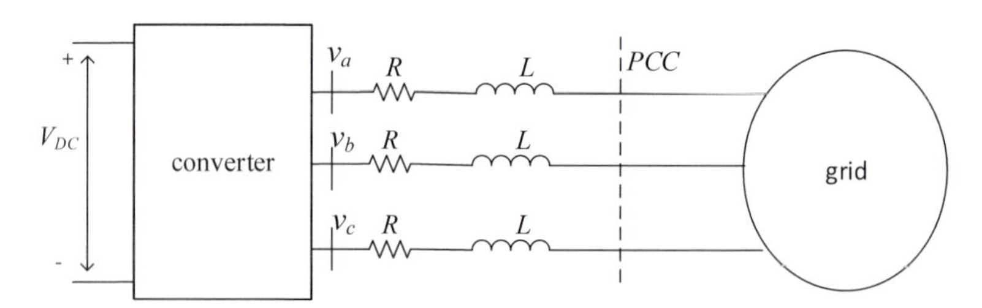

Figure 6.1: A circuit diagram of a converter connected to a grid.

The *dq*-reference frame is now utilized. It is assumed that the *d*-axis is aligned with the space vector of the PCC voltage.

$$L\frac{d(i_d + ji_q)}{dt} + j\omega L(i_d + ji_q) + R(i_d + ji_q) \\ = v_d + jv_q - (v_{1d} + jv_{1q}). \tag{6.2}$$

where ω is the grid's frequency, $v_{1q} = 0$ since the PCC voltage is aligned with the d-axis.

Separating equation (6.2) into dq-axes, the plant model for the current control design is derived:

$$L\frac{di_d}{dt} + Ri_d = \underbrace{v_d - v_{1d} + \omega L i_q}_{u_d} \tag{6.3}$$

$$L\frac{di_q}{dt} + Ri_q = \underbrace{v_q - v_{1q} - \omega L i_d}_{u_q} \tag{6.4}$$

The plant model for the current controller is assumed as $1/(R + sL)$ for both d and q axes. The inputs are u_d and u_q while the outputs are i_d and i_q. The feedback controls are designed for the dq-axis currents to track the reference signals. In addition, to generate the dq components of the converter voltage, the cross coupling and feed-forward voltage terms should be added after u_d and u_q are obtained from the controllers.

A simplified inner current control block is illustrated in Figure 6.2. The

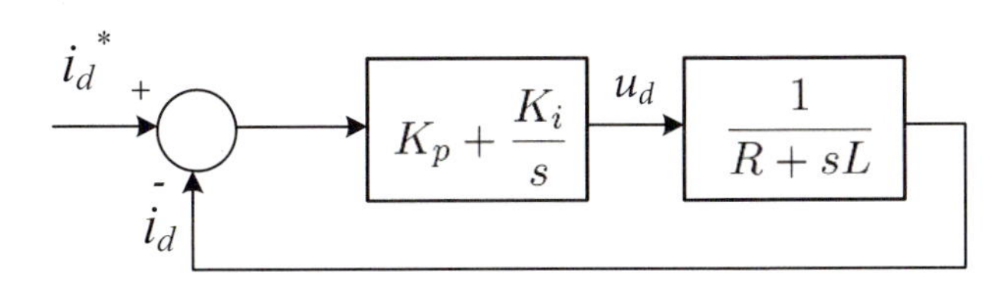

Figure 6.2: Simplified block diagram for inner loop control.

loop gain of system is represented by:

$$l(s) = \frac{K_p}{sL}\left(\frac{s + \frac{K_i}{K_p}}{s + \frac{R}{L}}\right). \tag{6.5}$$

As mentioned in Yazdani and Iravani (2010), the plant pole is fairly close to the origin. Therefore, this plant pole is canceled by the compensator's zero ($\frac{K_i}{K_p} = \frac{R}{L}$) and the loop gain becomes:

$$l(s) = \frac{K_p}{sL} \tag{6.6}$$

The closed-loop transfer function can be represented as:

$$G_{\text{Inner}}(s) = \frac{l(s)}{1 + l(s)} = \frac{1}{\tau s + 1} \tag{6.7}$$

where $\tau = \frac{L}{K_p}$ and $K_i = \frac{R}{\tau}$.

If $R = 0.02 \quad \Omega$ and $L = 0.04$ H, the inner loop with $K_i = 50$, $K_p = 100$ will lead to a bandwidth around 2500 rad/s ($\tau = 0.4$ ms).

6.1.2 Phase-Locked Loop (PLL)

With the inner current controller, u_d and u_q will be produced. The converter dq-axis reference voltage v_d^* and v_q^* can then be computed as

$$\begin{aligned} v_d^* &= u_d + v_{1d} - \omega L i_q \\ v_q^* &= u_q + v_{1q} + \omega L i_d \end{aligned} \tag{6.8}$$

Note that the generation of v_d^* and v_q^* requires the feedforward units (v_{1d} and v_{1q}) from the PCC voltage and the cross coupling terms ($-\omega L i_q$ in d-axis and $\omega L i_d$ in the q-axis).

Further, since the inputs of the PWM of a VSC are *abc* voltages, the dq-axis voltages need to be converted to v_{abc}. Note that the dq-reference frame is based on the space vector of the PCC voltage and the d-axis is aligned with the PCC voltage space vector.

Therefore, the space vector of the converter voltage will be

$$\vec{v} = (v_d + j v_q) e^{j\theta} \tag{6.9}$$

where θ is the PCC voltage space vector's angle relative to the static reference frame.

v_a, v_b, and v_c can be found from the space vector.

$$\begin{cases} v_a = \Re(\vec{v}) - v_d \cos\theta - v_q \sin\theta \\ v_b = \Re(\vec{v} e^{-j\frac{2\pi}{3}}) = v_d \cos\left(\theta - \frac{2\pi}{3}\right) - v_q \sin\left(\theta - \frac{2\pi}{3}\right) \\ v_c = \Re(\vec{v} e^{j\frac{2\pi}{3}}) = v_d \cos\left(\theta + \frac{2\pi}{3}\right) - v_q \sin\left(\theta + \frac{2\pi}{3}\right) \end{cases} \tag{6.10}$$

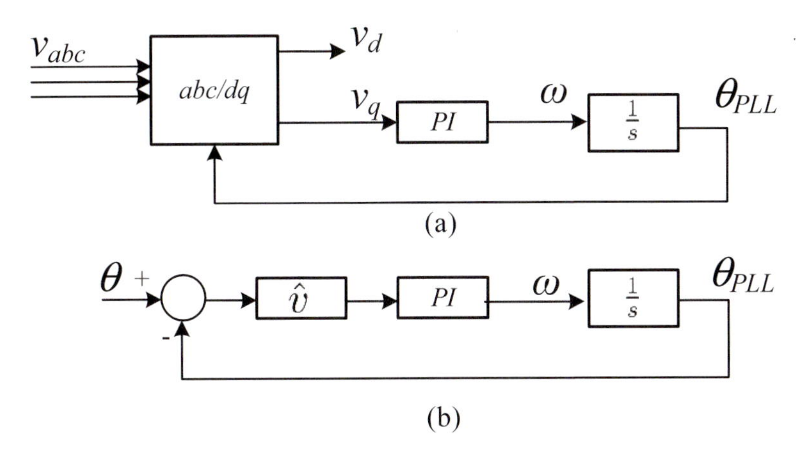

Figure 6.3: (a) Block diagram for a PLL. (b) Linearized block diagram for a PLL. The inputs of a PLL are three-phase voltages v_{abc}. The outputs of a PLL includes voltage magnitude v_d, frequency ω and angle θ_{PLL}.

The angle for the space vector $\vec{v_1}$ will be obtained by a PLL. Figure 6.3 presents the control diagram for a simple PLL.

The inputs to a PLL are three-phase voltages. For the application of a grid-connected VSC, the voltages are the PCC voltages v_{1a}, v_{1b} and v_{1c}. The voltages will be converted to dq-axis voltages based on an angle θ_{PLL}. A feedback loop equipped with integral controls now tries to make the q-axis voltage zero. This way, the input voltage's space vector is now aligned to the d-axis of a reference frame based on θ_{PLL}.

PLL is designed to have a bandwidth at 100 Hz. To conduct the design, we again first sought the linearized control blocks.

Using the space vector concept, we can show that the first block abc to dq is in fact the following input-output relationship:

$$v_d + jv_q = \vec{v}e^{-j\theta_{PLL}} = \hat{v}e^{j\theta}e^{-j\theta_{PLL}} = \hat{v}e^{j(\theta-\theta_{PLL})} \tag{6.11}$$

Therefore

$$\begin{cases} v_d &= \hat{v}\cos(\theta - \theta_{PLL}) \\ v_q &= \hat{v}\sin(\theta - \theta_{PLL}) \end{cases} \tag{6.12}$$

Assuming that $\theta - \theta_{PLL} \approx 0$, $\hat{v}$ is constant, the linearized model of v_q can now be expressed as

$$v_q = -\hat{v}(\theta - \theta_{PLL}) \tag{6.13}$$

Assuming that $\hat{v} \approx 1$, the closed-loop system in Figure 6.3(b) is expressed as

$$\frac{\theta_{PLL}}{\theta} = \frac{K_p s + K_i}{s^2 + K_p s + K_i}. \tag{6.14}$$

If $K_p \ll K_i$, then the system's bandwidth is $\sqrt{K_i}$. To obtain a 100 Hz bandwidth, we will select the parameters as follows: $K_i = (2\pi \times 100)^2 = 3.95 \times 10^5$. To have a 10% damping ratio, we will select $K_p = 2 \times 0.1 \times \sqrt{K_i} = 125.66$.

The Bode plot of the closed-loop system is shown in Figure 6.4. The Bode plot confirms that the peak occurs at 100 Hz. The bandwidth is approximately at 100 Hz.

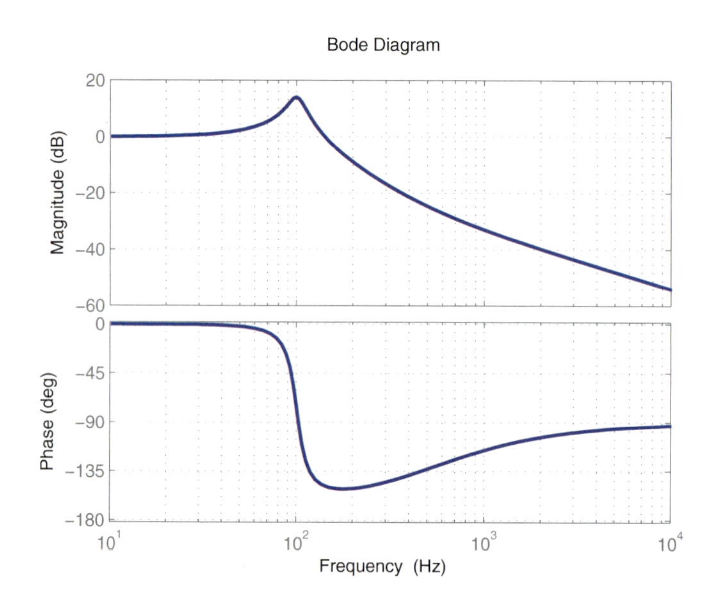

Figure 6.4: Bode diagram of the system of (6.14).

A Simulink model is built and shown in Figure 6.5. The simulation results are shown in Figure 6.6.

The abc/dq block is built based on the following. v_d and v_q can be found from v_{abc}. The space vector $\overrightarrow{v}$ will be expressed by v_a, v_b, and v_c.

$$\begin{aligned} v_d + jv_q &= \overrightarrow{v} e^{-j\theta_{PLL}} \\ &= \left(v_a + v_b e^{j\frac{2\pi}{3}} + v_c e^{-j\frac{2\pi}{3}}\right) e^{-j\theta_{PLL}} \end{aligned} \tag{6.15}$$

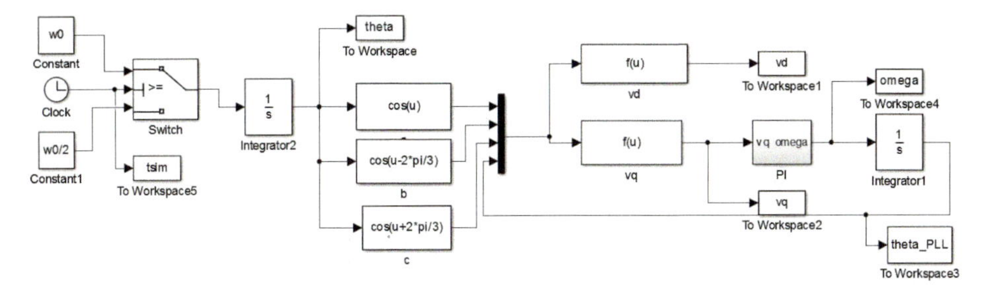

Figure 6.5: Simulink block diagram of a PLL test system.

Therefore,

$$\begin{cases} v_d = \frac{2}{3}\left(v_a \cos(\theta_{PLL}) + v_b \cos\left(\theta_{PLL} - \frac{2\pi}{3}\right) + v_c \cos\left(\theta_{PLL} + \frac{2\pi}{3}\right)\right) \\ v_q = -\frac{2}{3}\left(v_a \sin(\theta_{PLL}) + v_b \sin\left(\theta_{PLL} - \frac{2\pi}{3}\right) + v_c \sin\left(\theta_{PLL} + \frac{2\pi}{3}\right)\right) \end{cases} \tag{6.16}$$

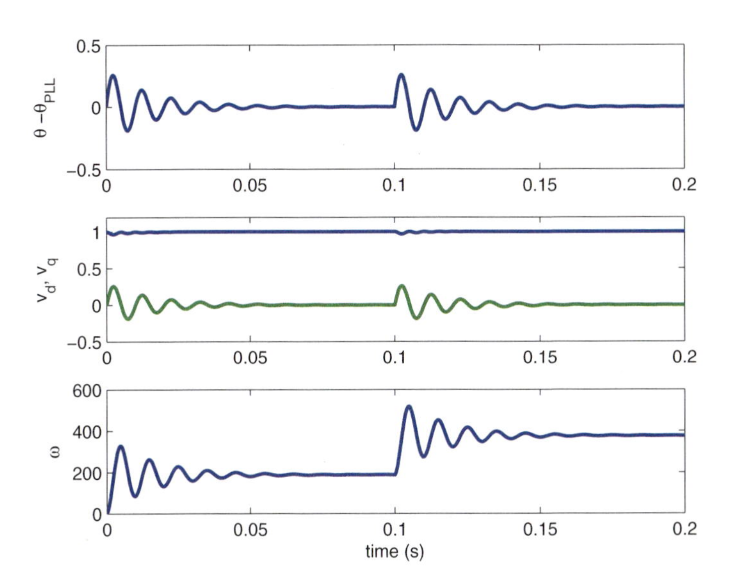

Figure 6.6: Simulation results of Figure 6.5. At $t = 0.1$s, the frequency of the input signal increases from 188.5 rad/s to 377 rad/s.

The simulation results show that when there is a change of frequency in the input signals, the PLL is able to correctly identify the frequency within

0.05 seconds. The output angle of the PLL is able to track the input signal's angle.

The overall current control structure with a PLL is shown in Figure 6.7, which includes inner current control, a PLL, and outer PQ control. This type of control is suitable for grid integration. The interfaced distributed energy resource (DER) generates or absorbs scheduled power.

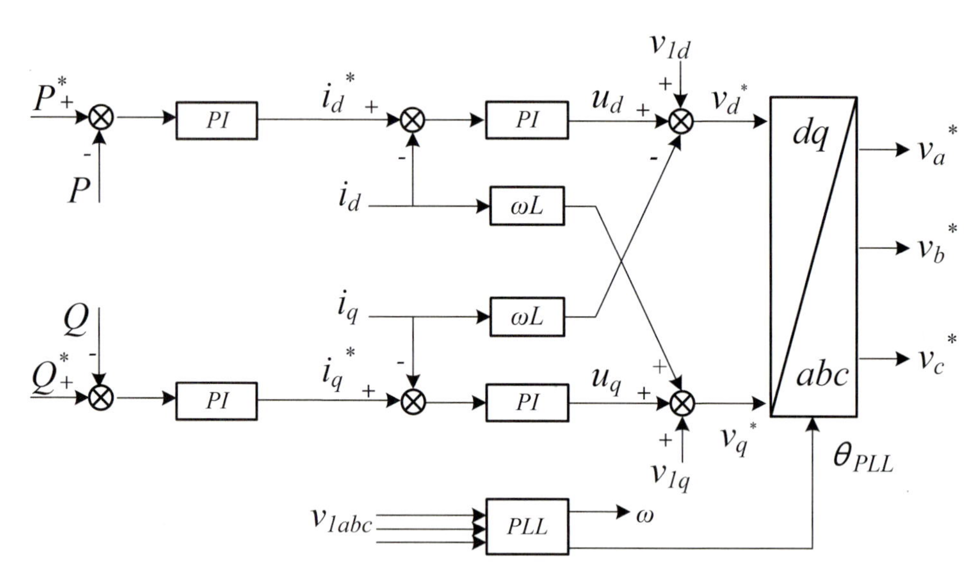

Figure 6.7: A VSC control structure. The control consists of inner current controls and outer power controls. PI stands for proportional integral control while PLL stands for phase-locked loop.

6.1.3 Validation of current control and PLL

After the design of the current controllers and a PLL, a MATLAB/Simulink model is built to validate their performance. The validation testbed consists of the plant model (the RL circuit), the PLL, and the current controllers. The testbed is shown in Figure 6.8.

The three embedded MATLAB functions used in Simulink are to convert v_{dq} to v_{abc}, compute i_{abc} given the converter voltage v_{abc} and the grid voltage v_{1abc}, and converter i_{abc} to i_{dq}. For each abc/dq or dq/abc conversion block, the PCC voltage angle has to be used as the input. The angle of the PCC voltage is measured by the PLL block. The inputs of the PLL block are the *abc* voltages of the grid.

The codes of the three blocks are shown as follows.

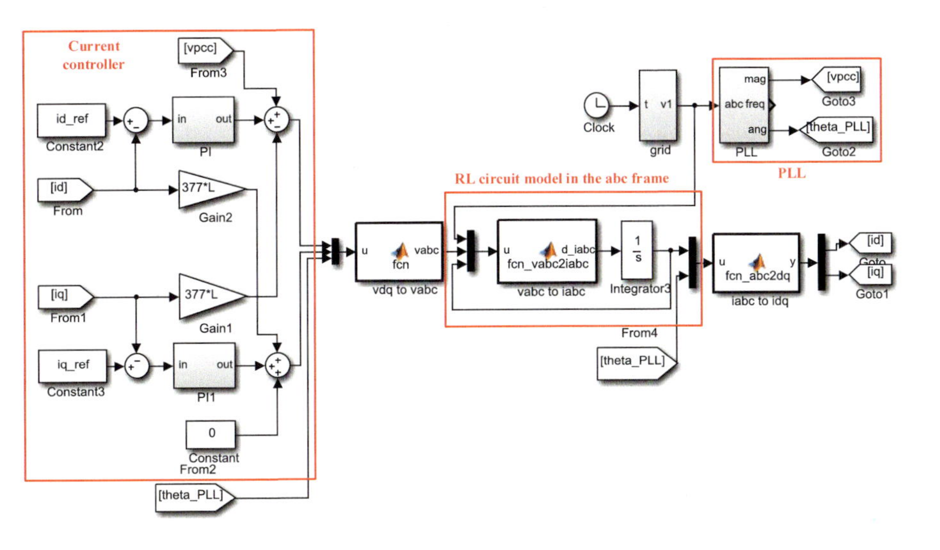

Figure 6.8: Simulation test bed to validate the performance of a system with current controllers and PLL. $R = 0.02\Omega$, $L = 0.04$ H. Current controller: $K_p = 50$, $K_i = 100$. PLL: $K_p = 125.66$, $K_i = 3.95 \times 10^5$.

Convert v_{dq} to v_{abc} given the inputs as v_{abc} and θ_{PLL}.

```
function vabc = fcn(u)
%#codegen
dq = u(1:2);
theta = u(3);
A =[cos(theta), -sin(theta);
    cos(theta-2*pi/3), -sin(theta-2*pi/3);
    cos(theta+2*pi/3), -sin(theta+2*pi/3)];
vabc = A*dq;
```

Compute $\frac{di_{abc}}{dt}$ given the inputs as v_{1abc}, v_{abc}. R and L are treated as parameters.

```
function d_iabc = fcn_vabc2iabc(u, R, L)
%#codegen
v1a = u(1); v1b = u(2); v1c = u(3);
va = u(4); vb = u(5); vc = u(6);
ia = u(7); ib = u(8); ic = u(9);
d_iabc = 1/L*[va- v1a - R*ia; vb-v1b-R*ib; vc-v1c-R*ic];
```

Compute i_{dq} given i_{abc} and θ_{PLL}.

```
 function y = fcn_abc2dq(u)
%#codegen
```

```
abc = u(1:3); theta = u(4);
A =2/3*[cos(theta), cos(theta-2*pi/3),  cos(theta+2*pi/3);
 -sin(theta), -sin(theta-2*pi/3),  -sin(theta+2*pi/3)];
y = A*abc;
```

Note that the reference dq-axis currents are given as constants. All initial values of integrators are set as zero. Figure 6.9 shows simulation results for this testbed.

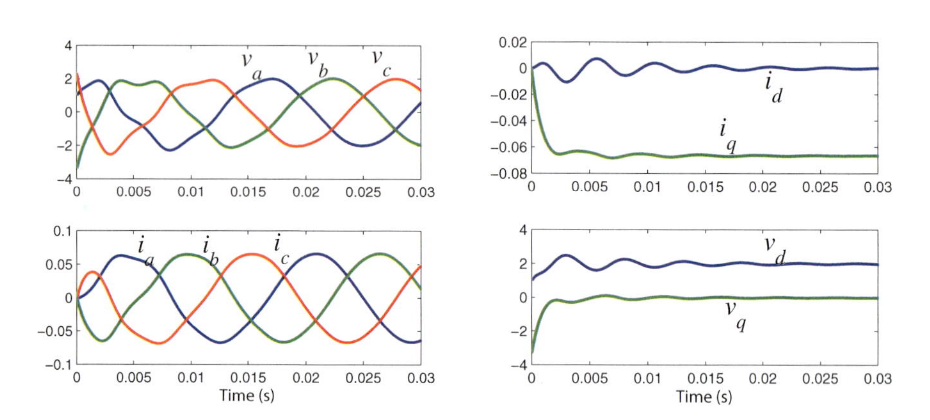

Figure 6.9: Simulation tests of the current control blocks and PLL.

It can be seen that within 0.03 seconds, the dq-currents reach steady-state values (the reference values).

6.1.4 Design of outer PQ or PV control

Compared to the inner current controller, the outer loop is designed to be very slow to reflect the dynamic changes. The simplified block diagram of the outer control loop is illustrated in Figure 6.10. As the d-axis of the dq reference frame is aligned with the PCC voltage, the per unit real and reactive powers can be expressed as $P = V_1 i_d$ and $Q = V_1 i_q$, where V_1 is the magnitude of the per-phase voltage.

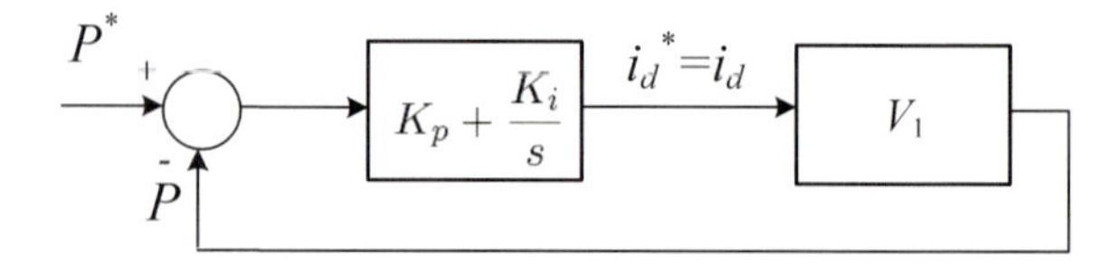

Figure 6.10: Simplified block diagram for outer loop control.

A closed-loop simplified transfer function is represented as:

$$G_{\text{outer}}(s) = \frac{\left(k_p + \frac{k_i}{s}\right) V_1}{1 + \left(k_p + \frac{k_i}{s}\right) V_1} = \frac{\frac{k_p}{k_i} s + 1}{\left(\frac{1}{V_1 k_i} + \frac{k_p}{k_i}\right) s + 1} \tag{6.17}$$

This is a first-order transfer function to the form of $\frac{as+1}{\tau s+1}$, where τ is the time constant ($\tau = \left(\frac{1}{V_1 k_i} + \frac{k_p}{k_i}\right)$) and the system bandwidth can be found as $1/\tau$. In this study, the outer loop gains are designed so that the bandwidth of the outer loop with $k_p = 0.1$ and $k_i = 5$ is 4.5 rad/s. This bandwidth is 300 times slower than the inner control bandwidth.

AC Voltage Control Ac voltage control is often used to replace Q control. The objective of the ac voltage control is to maintain the PCC bus voltage to a reference value. To validate the design, the PCC and the grid should be connected through a line. If the PCC voltage is the grid voltage, there will be no need to control the PCC voltage. The ac voltage PI controller is designed based on $\Delta Q = V_1 \Delta i_q$, where V_1 is the PCC voltage. Furthermore, the PCC voltage change ΔV_1 is proportional to ΔQ. Hence, the plant model is derived as:

$$\Delta V_1 \approx \frac{\Delta Q}{S_{sc}} = \frac{V_1}{S_{sc}} \Delta i_q \tag{6.18}$$

where S_{sc} notates the short-circuit capacity.

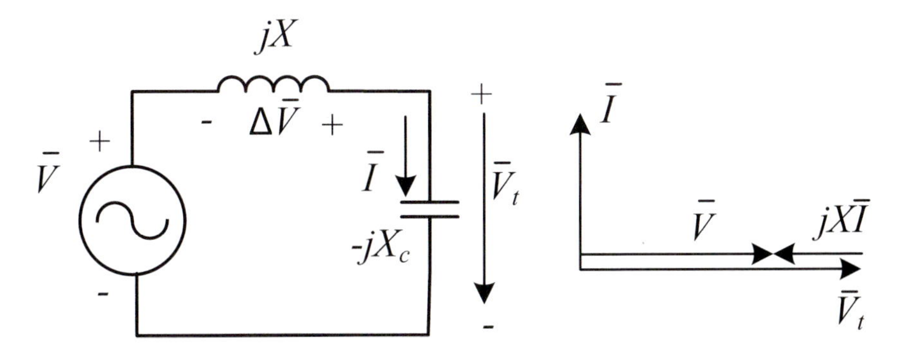

Figure 6.11: Effect of shunt reactive power compensation.

The voltage variation and the reactive power injection relationship is illustrated in Figure 6.11 and explained briefly here.

If a system can be represented by a Thevenin equivalent circuit with a voltage V behind a reactance X, then the short-circuit capacity can be found by shorting the two terminals: $S_{sc} = |S| = V \times \frac{V}{X}$. Assuming that $V \approx 1$ pu, then we have $S_{sc} = \frac{1}{X}$ where X is also in pu.

If we consider a shunt capacitor with reactance X_c (reactive power supported at the nominal voltage is $Q = \frac{1}{X_c}$) is connected to the terminal. Then the voltage increase will be:

$$\begin{aligned}
\Delta\overline{V} &= \overline{V}_t - \overline{V} = -jX\overline{I} \\
&= -jX\frac{\overline{V}_t}{-jX_c} \\
&\approx \frac{X}{X_c}\overline{V} \quad \text{assume that the votlage variation is small} \\
&= \frac{Q}{S_{sc}}
\end{aligned} \tag{6.19}$$

The transfer function of the closed-loop system can be found as:

$$\frac{\Delta V_1}{\Delta V_1^{\text{ref}}} = \frac{k_p + \frac{k_i}{s}}{1 + \left(k_p + \frac{k_i}{s}\right)\frac{1}{S_{sc}}} = \frac{\frac{k_p}{k_i}s + 1}{\frac{k_p + SCR}{k_i}s + 1}S_{sc} \tag{6.20}$$

with the assumption that V_1 is approximately 1 pu. Therefore, the time constant and the bandwidth are as follows:

$$\tau = \frac{k_p + S_{sc}}{k_i} \tag{6.21}$$

$$\omega_{bw} = 1/\tau \tag{6.22}$$

For $k_p = 0.01$, $k_i = 100$ and $S_{sc} = 1$, the bandwidth is 100 rad/s. For $S_{sc} = 2$, the bandwidth is 50 rad/s.

Validation of PQ Control

The testbed shown in Figure 6.8 will be modified to include a PQ controller. Figure 6.12 shows the simulation results. At $t = 0.25$ s, the reference power P^* changes from 0 to 0.1 pu. At $t = 1$ s, the reference reactive power Q^* changes from 0 to 0.1 pu. Figure 6.12 shows that the PQ controller can make the converter's output real power and reactive power track the reference values. A larger integral gain k_i results in a faster response.

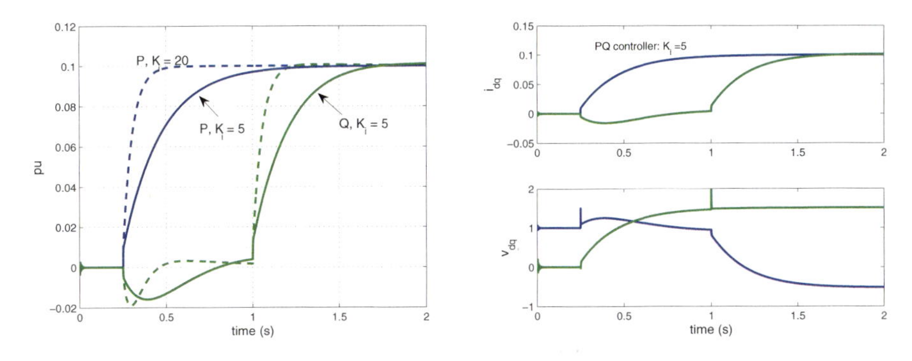

Figure 6.12: Simulation tests of the PQ control blocks. PQ controller: $k_p = 0.1$.

6.1.5 Design of VF control

For autonomous operation, a converter can be operated in the VF control mode. The objective is to (i) maintain the PCC voltage at a reference value, and (ii) maintain the system frequency at a reference value. For a system of a VSC serving a load is shown in Figure 6.13, the PCC voltage is the load voltage or the capacitor voltage e.

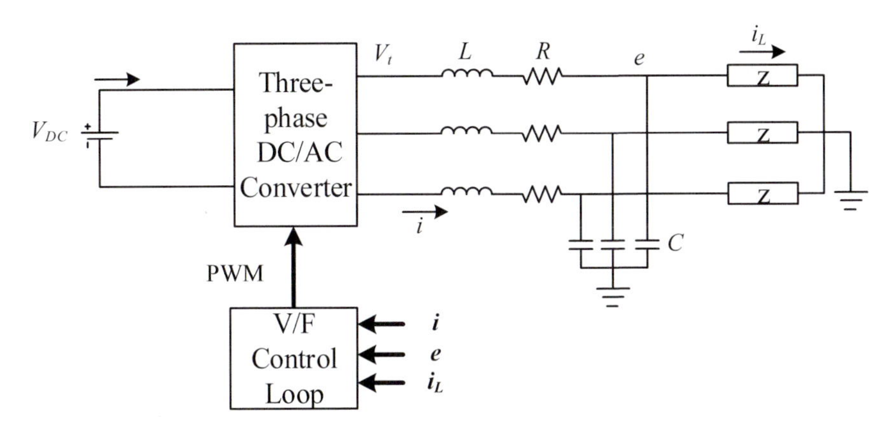

Figure 6.13: An autonomous microgrid with a VSC serving a load.

The voltage to be controlled is the capacitor voltage. In both VF and PQ control modes, the inner current control loops will always be there. Therefore, in PQ control mode, the outer PQ control determines the reference currents. Here in the VF mode, the outer VF control determines the reference currents as well. We will first set up the plant model. Since the controller's inputs are the capacitor voltages and its outputs are reference currents, the plant model should have the current as the input and the capacitor voltage as the output. For simplicity, we again use the assumption that the dynamics of the inner loop is much faster than the dynamics of the outer loop; therefore, we can consider $i_d^* \approx i_d$ and $i_q^* \approx i_q$ from the perspective of the outer loop.

The dynamics of the capacitor can be described in the *abc* frame as well as in the *dq* frame.

$$C\frac{de}{dt} = i - i_L, \quad abc \text{ frame}$$

$$C\frac{d(e_d + je_q)}{dt} + j\omega C(e_d + je_q) = (i_d + ji_q) - (i_{Ld} + ji_{Lq}), \quad dq \text{ frame}$$

Separating the real and imaginary elements of the equation in the *dq*-frame, we have equations in *d*-axis and *q*-axis.

$$C\frac{de_d}{dt} - \omega C e_q = i_d - i_{Ld} \tag{6.23}$$

$$C\frac{de_q}{dt} + \omega C e_d = i_q - i_{Lq} \tag{6.24}$$

The plant models suitable for linear control system design are the following:

$$\frac{e_d}{u_d} = \frac{e_q}{u_q} = \frac{1}{Cs} \tag{6.25}$$

where

$$u_d = i_d - i_{Ld} + \omega C e_q \tag{6.26}$$

$$u_q = i_q - i_{Lq} - \omega C e_d \tag{6.27}$$

Based on the simple first-order plant model, we can design PI controllers to track the reference voltage values e_d^* and e_q^*. The outputs of the PI controllers are u_d and u_q. We will find the reference currents i_d^* and i_q^* using the following relationships.

$$i_d^* \approx i_d = u_d + i_{Ld} - \omega C e_q \tag{6.28}$$

$$i_q^* \approx i_q = u_q + i_{Lq} + \omega C e_d \tag{6.29}$$

The frequency of the three-phase *abc* voltage v_{tabc} can be fixed at the nominal. This is done by feeding the dq/abc conversion block an angle $\omega_0 t$ where ω_0 is the nominal frequency.

The complete VF control block diagram is shown in Figure 6.14.

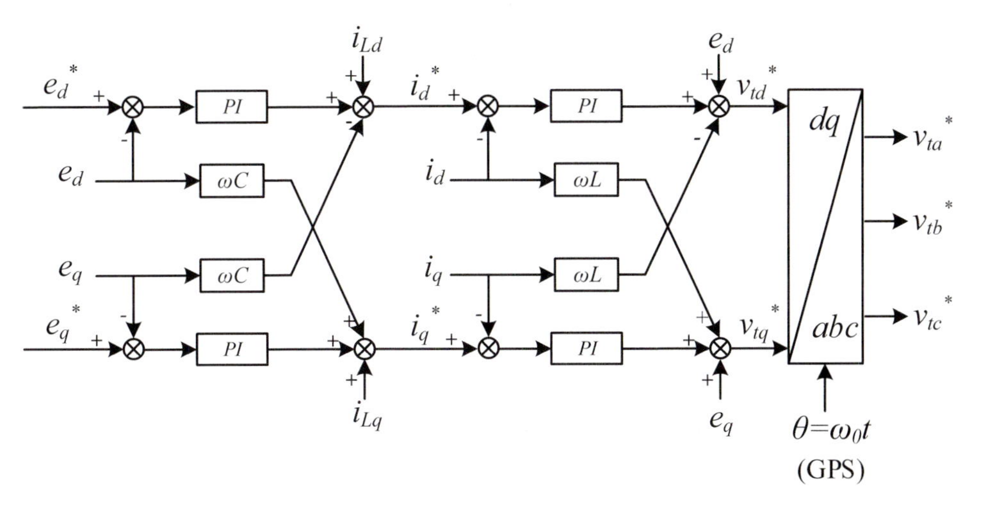

Figure 6.14: VF control.

Validation of the VF control

A MATLAB/Simulink model for the system shown in Figure 6.13 is built to validate VF control performance. The Simulink block diagram is shown in Figure 6.15. This block diagram consists of mainly two parts: the circuit dynamics and the controls. Outputs of the circuit dynamics blocks are measurements, such as converter currents i_d, i_q, load currents i_{Ld} and i_{Lq}. These measurements will be used in the control blocks.

Outputs of the control blocks are the converter voltage v_d and v_q. In addition, dq/abc and abc/dq transformation blocks are necessary if we model the circuit dynamics in the abc frame.

The circuit dynamics are modeled in the abc frame. There are three state variables for each phase, the capacitor voltage e_k, the converter current i_k, and the load current i_{Lk}, where $k = a, b, c$. The differential equations related to these state variables are shown as follows. The load is assumed to be an RL load and the load resistance is R_L while the load inductance is L_L.

$$\begin{cases} C\frac{de_k}{dt} & = i_k - i_{Lk} \\ L\frac{di_k}{dt} & = v_{tk} - e_k - Ri_k \\ L_L\frac{di_{Lk}}{dt} & = e_k - R_L i_{Lk} \end{cases} \tag{6.30}$$

In a MATLAB/Simulink block diagram, the state variables are the vector outputs from the integrator. The block feeding the integrator computes the derivatives of the state variables. This block has inputs from two sources:

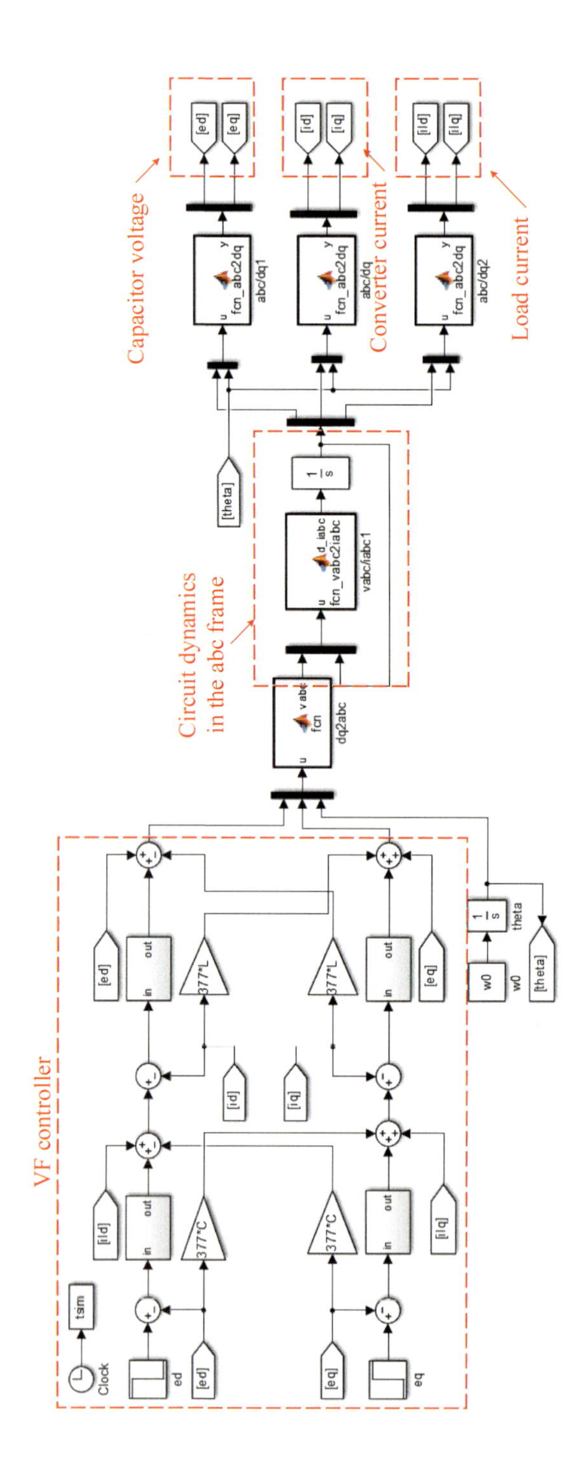

Figure 6.15: A single VSC serving load system with VF control. Note that the angle input to the dq/abc block is generated from a constant frequency integrated over time. Therefore, $\theta = \omega_0 t$. The parameters are as follows. $R = 0.02$, $L = 0.04$, $R_L = 3$, $L_L = 0.01$. Current PI controllers: $50 + 100/s$. Voltage PI controllers: $8 + 10/s$.

v_{abc} from the converter control block, i_{abc} and i_{Labc} from the output of the integrator. The code of this block is shown as follows.

```
function d_iabc = fcn_vabc2iabc(u)
%#codegen
vta = u(1); vtb = u(2); vtc = u(3);
ea = u(4); eb = u(5); ec = u(6);
ia = u(7); ib = u(8); ic = u(9);
ila=u(10); ilb=u(11); ilc=u(12);

R = 0.02; L = 0.04; Rl=3; Ll=0.01;  C=0.001;
de=1/C*[(ia-ila);(ib-ilb);(ic-ilc)];
di=1/L*[vta-ea-R*ia;vtb-eb-R*ib;vtc-ec-R*ic];
dil=1/Ll*[ea-Rl*ila;eb-Rl*ilb;ec-Rl*ilc];
d_iabc = [de;di;dil];
```

Step responses of the dq-axis voltages are tested. Here, e_d^* is changed at $t = 0.05$ s while e_q^* is changed at $t = 0.1$ s, the simulation results are shown in Figure 6.16. The two voltages can track their references very well.

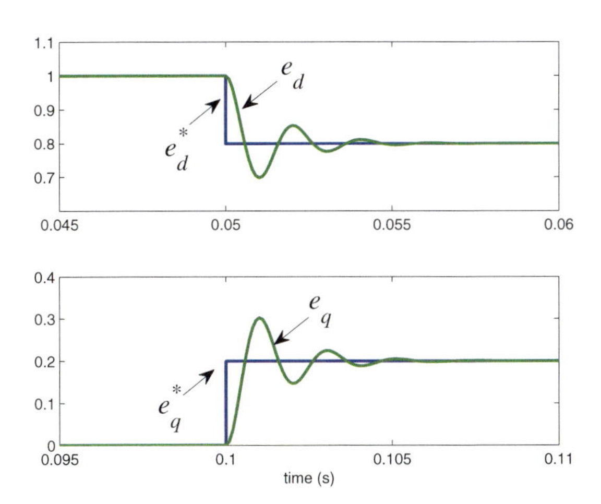

Figure 6.16: Step responses of the VF controller.

6.2 Power sharing methods

6.2.1 P–f and Q–E droops

In this section, power sharing methods among converters will be discussed. Droop controls that are applied in synchronous generators and reactive power sharing again find applications in converters.

The droop control equations are as follows.

$$\omega = \omega^* - m(P - P^*) \tag{6.31}$$
$$E = E^* - n(Q - Q^*) \tag{6.32}$$

where ω^* and E^* are the frequency and magnitude of the output voltage at this condition: $P = P^*$ and $Q = Q^*$.

However, implementation of droop control depends on the converter control structure. Section 6.1 shows two types of converter controls: PQ or VF. For each type, the implementation of droop control is different. Figure 6.17 presents the droop control implementation in PQ mode and VF mode.

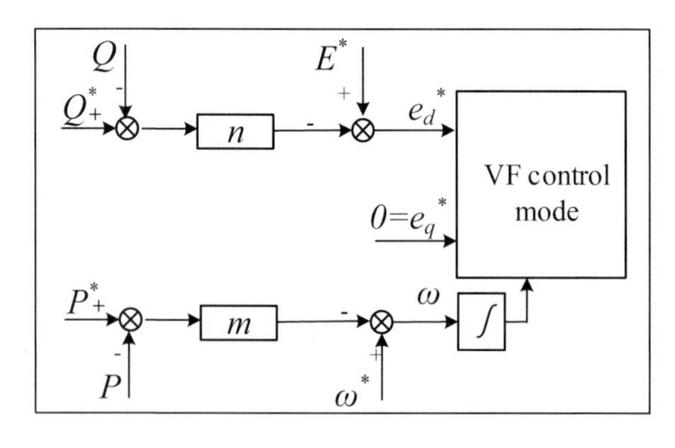

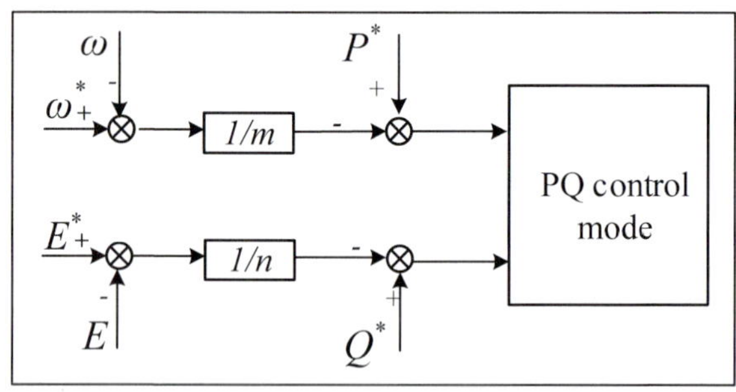

Figure 6.17: Two types of droop implementation.

Example of power sharing among two parallel converters in PQ control mode

In this example, we examine a system with two VSCs serving a load shown in Figure 6.18.

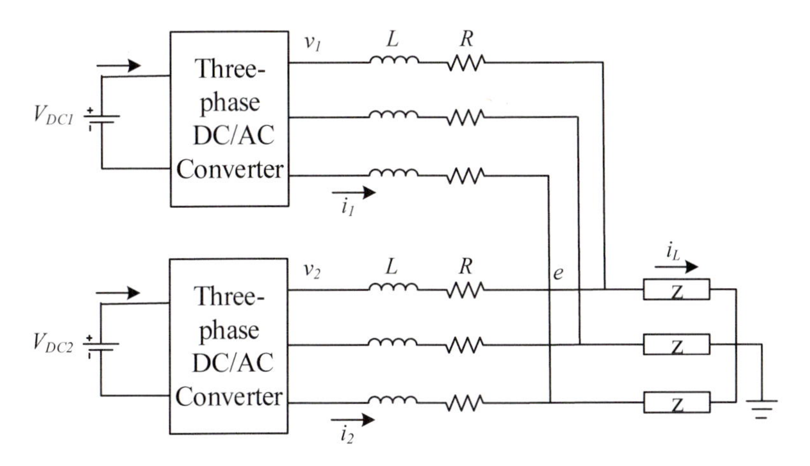

Figure 6.18: Two VSCs serving a resistive load. $R = 0.02$, $L = 0.04$, $R_L = 1.5$ before 0.6 seconds and $R_L = 1.2$ after 0.6 seconds.

The VSCs are all in PQ control mode. On top of the PQ control, P–f and Q–E droops are applied. The voltage measurement used in the Q–E droop is the PCC voltage while the frequency measurement used in the P–f droop is the PCC voltage frequency. A PLL is used to obtain the PCC voltage as well as its frequency.

The MATLAB/Simulink model is shown in Figure 6.19.

Note that at the initial condition, each VSC shares half of the real power 0.5 unit while the reactive power to the load are all zeros since the load is resistive. The droop coefficients $m_1 : m_2 = n_1 : n_2 = 1 : 2$, which indicates when there is a load change, VSC2 will take 2/3 of the share while the VSC1 will take 1/3 of the share.

The PCC voltage is kept at 1 pu. The decrease of the load resistance from 1.5 pu to 1.2 pu makes the real power consumption increase from 1 pu to 1.25 pu. For the 0.25 pu load change, 0.167 pu will be generated by VSC2 while 0.083 pu will be generated by VSC1. Hence the steady-state value of P_2 will be 0.667 pu while P_2 will be 0.583 pu. The simulation results are presented in Figure 6.20. Note that due to the load increase, the system frequency will be reduced. $\Delta\omega$ can be found from the droop coefficients. Since $\Delta P_L = (m_1 + m_2)\Delta\omega$, therefore, $\Delta\omega = \frac{0.25}{0.04+0.08} = 2.08$ rad/s. This result can be confirmed by the simulation result of ω.

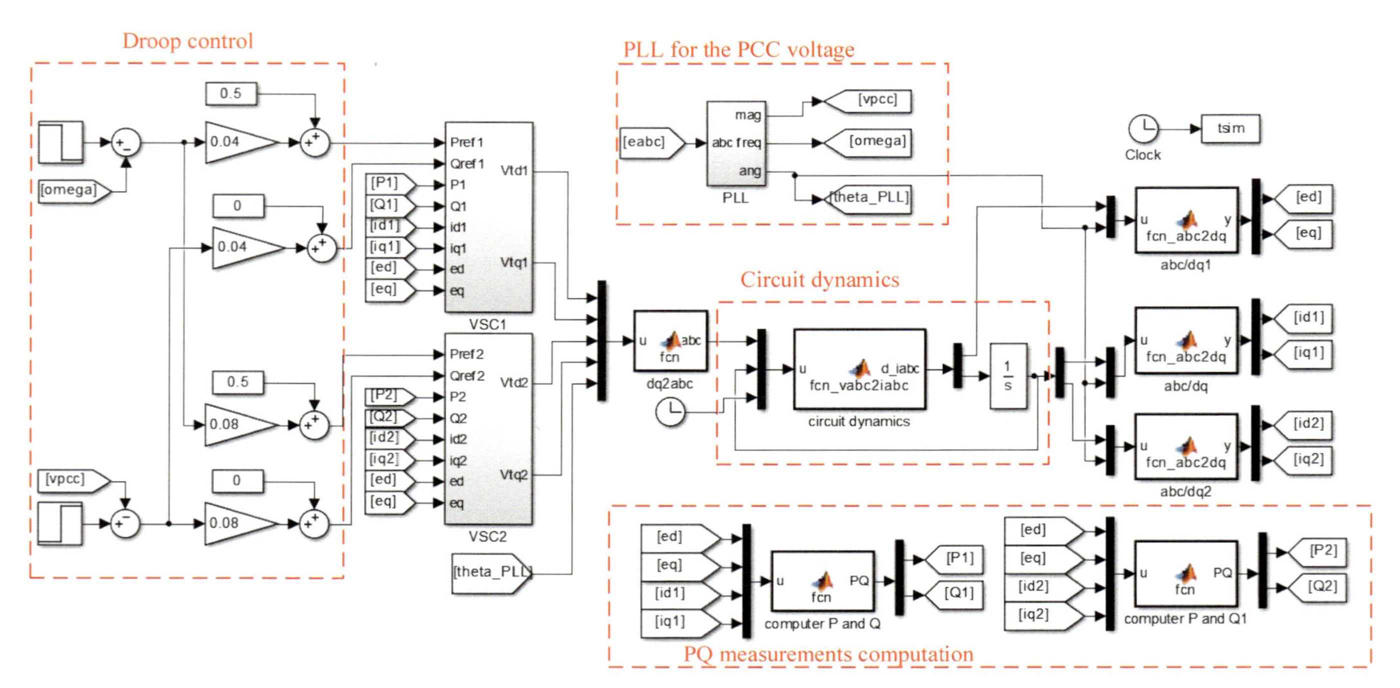

Figure 6.19: Two VSCs serving a load system with droop control. Each VSC is in PQ control mode.

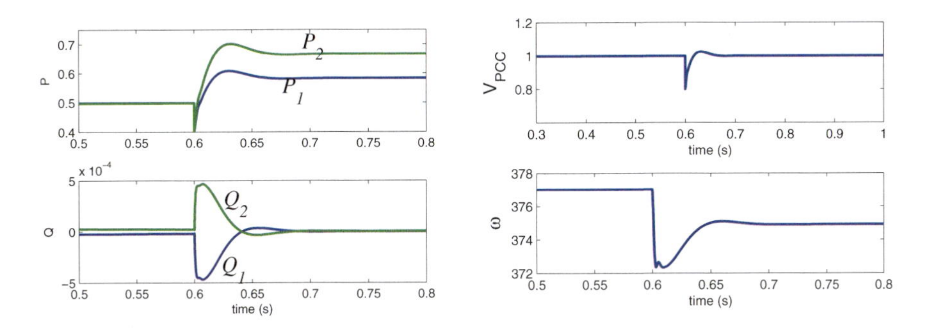

Figure 6.20: Simulation results for a load change at $t = 0.6$ s.

6.2.2 V–I droop

V–I droop is more aligned to the droop design in STATCOM for reactive power sharing. For converters interfacing distributed energy resources, not only reactive power sharing but also real power sharing should be designed. V–I droop has the capability for both real power and reactive power sharing.

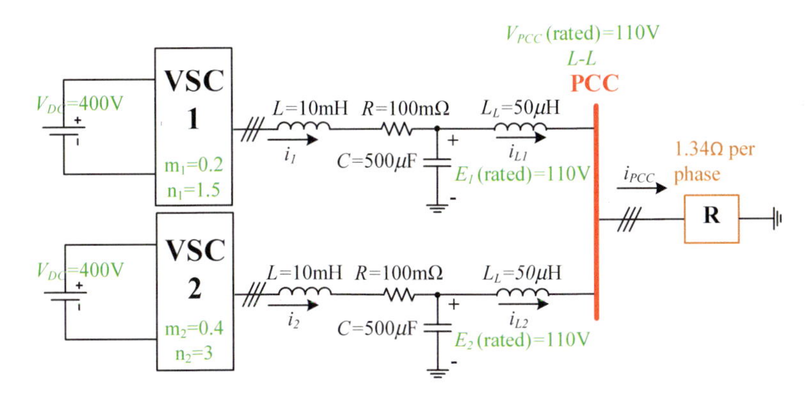

Figure 6.21: Two DERs support one load through parallel VSCs.

Assume that the two converters (including the RLC filter) are connected to the PCC bus through an RL impedance $R_k + jX_k, k = 1, 2$ (Figure 6.21. The complex power injected to the load or the grid through the PCC bus is notated as $S_k, k = 1, 2$. Let the d-axis be aligned with the PCC voltage space vector ($v_d = |\vec{v}_{PCC}|$). Then the complex power expression is

$$S_k = \frac{3}{2} v_d (i_{Ldk} - j i_{Lqk}) = \frac{3}{2} (v_d i_{Ldk} - j v_d i_{Lqk}) \tag{6.33}$$

Therefore,

$$P_k = \frac{3}{2} v_d i_{Ldk} \quad (6.34a)$$

$$Q_k = -\frac{3}{2} v_d i_{Lqk}. \quad (6.34b)$$

Therefore, the d-axis current sharing determines the real power sharing while the q-axis current sharing determines the reactive power sharing. At steady-state, the converter voltage, PCC voltage and the current have the following relationship.

$$\begin{bmatrix} R_k & -X_k \\ X_k & R_k \end{bmatrix} \begin{bmatrix} i_{Ldk} \\ i_{Lqk} \end{bmatrix} = \begin{bmatrix} E_{kd} \\ E_{kq} \end{bmatrix} - \begin{bmatrix} v_d \\ 0 \end{bmatrix} \quad (6.35)$$

If we apply the V–I droop and assume the droop coefficients are m_k for the d-axis and n_k for the q-axis, then we have the following relationship.

$$\begin{bmatrix} E_{kd} \\ E_{kq} \end{bmatrix} - \begin{bmatrix} v_d \\ 0 \end{bmatrix} = \begin{bmatrix} E_0 \\ 0 \end{bmatrix} - \begin{bmatrix} m_k & 0 \\ 0 & n_k \end{bmatrix} \begin{bmatrix} i_{Ldk} \\ i_{Lqk} \end{bmatrix} - \begin{bmatrix} v_d \\ 0 \end{bmatrix} \quad (6.36)$$

$$\Longrightarrow \begin{bmatrix} R_k + m_k & -X_k \\ X_k & R_k + n_k \end{bmatrix} \begin{bmatrix} i_{Ldk} \\ i_{Lqk} \end{bmatrix} = \begin{bmatrix} E_0 \\ 0 \end{bmatrix} - \begin{bmatrix} v_d \\ 0 \end{bmatrix} \quad (6.37)$$

If we assume that $m_k \gg R_k$, $n_k \gg R_k$, $m_k \gg X_k$, and $n_k \gg X_k$, then we have the following relationship:

$$m_1 i_{Ld1} = m_2 i_{Ld2} \quad (6.38a)$$

$$n_1 i_{Lq1} = n_2 i_{Lq2} \quad (6.38b)$$

Therefore, the real power sharing is proportional to $1/m_k$ and the reactive power sharing is proportional to $1/n_k$. The assumptions are $m_k \gg R_k$, $n_k \gg R_k$, $m_k \gg X_k$, and $n_k \gg X_k$. If we have a resistive network or $R_k \gg X_k$, then the real power sharing is according to $1/(R_k + m_k)$ and the reactive power sharing is according to $1/(R_k + n_k)$.

If the droop coefficients m_k and n_k are comparable with R_k or X_k, then we cannot obtain accurate real/reactive power sharing. On the other hand, large droop coefficients may lead to converter voltages drop below the range during heavy load conditions.

The control block diagram for the V–I droop is shown in Figure 6.22. Note that the droop is added on top of a VF controller.

For the system shown in Figure 6.21, the circuit dynamics and the controls are built in a MATLAB/Simulink model. The dynamic event checked

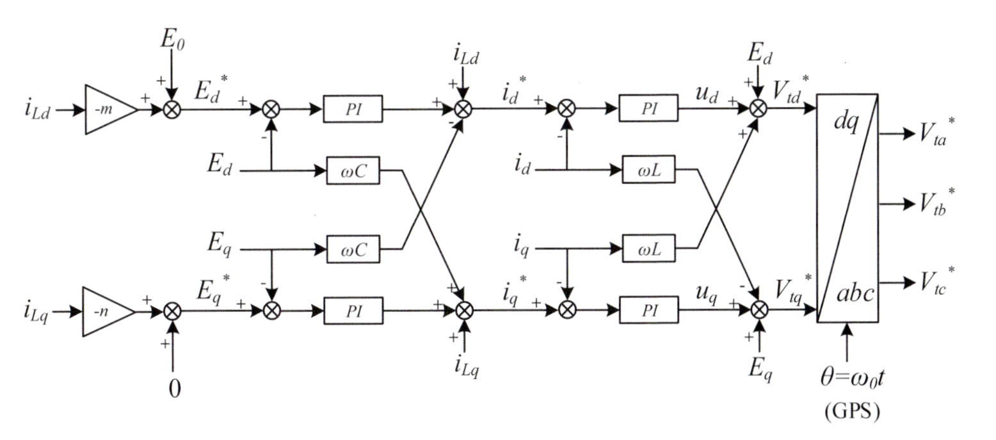

Figure 6.22: Control block diagram for V-I droop.

is a step change in the nominal PCC voltage E_0. The simulation results are shown in Figure 6.23. The parameters adopted are: $m_1 : m_2 = 0.2 : 0.4$ and $n_1 : n_2 = 1.5 : 3.0$. VSC1 is expected to share twice as much as VSC2 on active power. Since the load is resistive, the reactive power is very small for both VSCs. The simulation results in Figure 6.23 confirm the expectation.

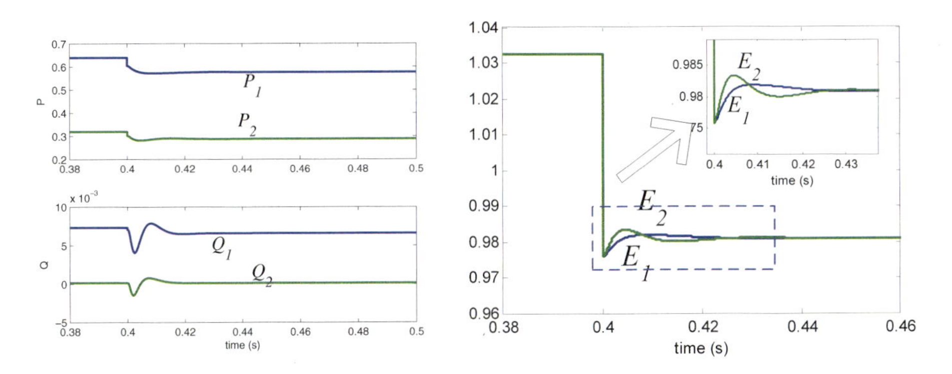

Figure 6.23: Simulation results for a step change in E_0.

Part II: Dynamics

Chapter 7

Large-Signal Stability

7.1 Introduction

In Part I, control design is based on linearized system models. The underlying assumption is that for small-signal disturbances, e.g., a small change in an automatic voltage regulator's voltage reference, we can use linear system analysis to conduct control design as well as stability analysis.

This is not the case with large-signal disturbances, for example, when a power grid is subject to a three-phase fault. Can we determine if the system will remain stable after the fault is cleared? This type of stability cannot be analyzed using linear system models and linear system analysis tools.

For large-signal disturbances, time-domain simulation is suitable to be used to examine the system behavior. For a simple system, we can also rely on Lyapunov stability theory to examine large-signal stability.

In the rest of this chapter, the Lyapunov stability criterion will be presented first. Then a SMIB system's large-signal stability, or transient stability will be examined by the equal-area method, which is based on the Lyapunov stability criterion. Analysis results are then validated by time-domain simulation results.

7.2 Lyapunov stability criterion

The Lyapunov stability theorem states that if the dynamics of a system can be described as

$$\dot{x} = f(x) \tag{7.1}$$

where $x \in \mathbb{R}^n$ and $f : \mathbb{R}^n \to \mathbb{R}^n$, then the system is stable if and only of there is a value function $V(x) \geq 0$ where $V : \mathbb{R}^n \to \mathbb{R}$ for all x on the dynamic trajectory, and its derivative is less than or equal to zero $\dot{V}(x) \leq 0$.

Example: Use the Lyapunov stability theorem to tell if the following system is stable or not.

$$\begin{aligned} \dot{x}_1 &= -x_1 + g(x_2) \\ \dot{x}_2 &= -x_2 + h(x_1) \end{aligned} \tag{7.2}$$

where

$$\begin{aligned} |g(u)| &\leq |u|/2 \\ |h(u)| &\leq |u|/2 \end{aligned}$$

Solution We can assume a value function $V(x) = \frac{1}{2}(x_1^2 + x_2^2)$. $V(x) \geq 0$ for all x. Next the derivative of $V(x)$ is evaluated.

$$\begin{aligned} \dot{V}(x) &= x_1\dot{x}_1 + x_2\dot{x}_2 = x_1(-x_1 + g(x_2)) + x_2(-x_2 + h(x_1)) \\ &= -x_1^2 - x_2^2 + x_1 g(x_2) + x_2 h(x_1) \\ &\leq -x_1^2 - x_2^2 + |x_1||g(x_2)| + |x_2||h(x_1)| \\ &\leq -x_1^2 - x_2^2 + |x_1|\frac{|x_2|}{2} + |x_2|\frac{|x_1|}{2} \\ &= -x_1^2 - x_2^2 + |x_1 x_2| \\ &\leq -\frac{1}{2}(x_1^2 + x_2^2) = -V(x) \end{aligned} \tag{7.3}$$

$\dot{V}(x) \leq 0$ for any x. Therefore, the above system is a stable system.

For a linear time invariant (LTI) system, the system dynamics can be described as

$$\dot{x} = Ax. \tag{7.4}$$

where $A \in \mathbb{R}^{n\times n}$ and $x \in \mathbb{R}^n$.

Assume that a value function $V(x) = x^T P x$, where $P \in \mathbb{R}^{n\times n}$ is a positive semi-definite (PSD) matrix, i.e., P is symmetric $P^T = P$ and every eigenvalue of P is greater than or equal to zero. We will notate it as $P \succeq 0$. Then $V(x) \geq 0$ for any x.

The above statement can be proved using eigenvalue decomposition. For P, we can decompose it to be $P = V^{-1}\Lambda V$ where Λ is a diagonal matrix with diagonal elements $\lambda_i \geq 0$ $(i = 1, \cdots, n)$ and V is the eigenvector matrix. V is orthogo-normal, i.e., $V^T = V^{-1}$.

$$V(x) = x^T V^T \Lambda V x = \tilde{x}^T \Lambda \tilde{x} = \sum_{i=1}^{n} \lambda_i \tilde{x}_i^2 \geq 0 \tag{7.5}$$

where $\tilde{x} = Vx$.

$$\dot{V}(x) = x^T P\dot{x} + \dot{x}^T Px = x^T(PA + A^T P)x \tag{7.6}$$

If the linear system is stable, then $\dot{V}(x) \leq 0$. This is equivalent to saying $PA + A^T P \preceq 0$.

7.2.1 Stability or instability

The Lyapunov stability theorem states that the system is stable if $V(x) \geq 0$ while $\dot{V}(x) \leq 0$ for all x on the trajectory. The system is unstable if $V(x) < 0$ while $\dot{V}(x) \leq 0$ for any x on the trajectory.

7.3 Equal-area method

We will use the Lyapunov stability theorem to judge if a SMIB power system is stable or not.

The dynamic equations for the system is described as follows.

$$\dot{\delta} = \omega_0(\omega - 1) \tag{7.7a}$$

$$\dot{\omega} = \frac{1}{2H}(P_m - P_e - D_1(\omega - 1)) \tag{7.7b}$$

We would like to construct a value function $V(\delta, \omega)$ and make sure its time derivative is less than or equal to zero for all δ, ω. If we pick

$$\begin{aligned}\dot{V}(\delta, \omega) &= -D_1\omega_0^2(\omega - 1)^2 \\ &= -D_1\dot{\delta}^2,\end{aligned} \tag{7.8}$$

then $\dot{V}(\delta, \omega) \leq 0$ for all (δ, ω). To judge if the system is stable or not, we just need to examine if $V(\delta, \omega) \geq 0$ for all (δ, ω) or $V(\delta, \omega) < 0$ for any (δ, ω) in the trajectory.

$$
\begin{aligned}
V(\delta,\omega) &= \int_{t_0}^{t} \dot{V}(\delta,\omega)dt = \int_{t_0}^{t} -D_1(\dot{\delta})^2 dt \\
&= -\int_{\delta_0}^{\delta} D_1\dot{\delta}d\delta \qquad \text{(Replace } D_1\dot{\delta} \text{ or } \omega_0 D_1(\omega-1) \text{ using (7.7b))} \\
&= -\int_{\delta_0}^{\delta} \left[\omega_0(P_m - P_e) - 2H\omega_0\dot{\omega}\right] d\delta \\
&= -\omega_0 \int_{\delta_0}^{\delta} (P_m - P_e)d\delta + \int_{\delta_0}^{\delta} 2H\frac{d\dot{\delta}}{dt}d\delta \\
&= -\omega_0 \int_{\delta_0}^{\delta} (P_m - P_e)d\delta + \int_{\delta_0}^{\delta} 2H\dot{\delta}d\dot{\delta} \\
&= \omega_0 \int_{\delta_0}^{\delta} (P_e - P_m)d\delta + H(\dot{\delta}^2)\Big|_{\delta_0}^{\delta}
\end{aligned} \tag{7.9}
$$

Note at the initial condition, $\dot{\delta} = 0$ when $\delta = \delta_0$. At a certain trajectory point when the speed reaches 1 pu or $\dot{\delta} = 0$, we only need to judge if $\int_{\delta_0}^{\delta}(P_e - P_m)d\delta < 0$. If so, then the system is unstable. This trajectory point corresponds to a maximum angle $\delta_{\max}$ or a minimum angle $\delta_{\min}$.

Therefore, equal area method computes

$$
\int_{\delta_0}^{\delta_{\max}} (P_e - P_m)d\delta
$$

if the rotor angle δ is increasing.

Example 1: For a SMIB system, the generator terminal bus is subjected to a three-phase to ground fault at t_0. the initial rotor angle is δ_0. The fault is cleared at $t_1 = t_0 + \Delta t$ and at t_1, the rotor angle is δ_1. Judge if the system will be stable or not. Assume that the mechanical power is kept constant. If the system is stable, what is the maximum $\delta_{\max}$?

Solution: We will examine $\int_{\delta_0}^{\delta_{\max}}(P_e - P_m)d\delta$. This expression can be separated into two components:

$$
\int_{\delta_0}^{\delta_{\max}} (P_e - P_m)d\delta = -\underbrace{\int_{\delta_0}^{\delta_1} (P_m - P_e)d\delta}_{A_1} + \underbrace{\int_{\delta_1}^{\delta_{\max}} (P_e - P_m)d\delta}_{A_2}. \tag{7.10}
$$

From t_0 to t_1, the short circuit fault is occurring and the terminal voltage of the generator is 0. In turn, the real power output from the generator

$P_e = 0$. Therefore, from t_0 to t_1, the net power imposed on the generator will cause the generator to accelerate. A_1 is called acceleration area. A_2 is called deceleration area since after t_1, the electric power P_e will be restored and $P_e > P_m$.

$$A_1 = P_m(\delta_1 - \delta_0)$$

The maximum of $\delta_{\max}$ can be is $\pi - \delta_0$. Since if $\delta > \delta_{\max}$, $P_e < P_m$ and the generator will accelerate. The accelerating area will be more than A_1. Therefore, the maximum A_2 is expressed as follows:

$$A_2 = \int_{\delta_1}^{\pi-\delta_0} \left(\frac{EV_\infty}{X} \sin(\delta) - P_m \right) d\delta$$

If $A_2 \geq A_1$, the system is stable. Otherwise, the system is not stable. Figure 7.1 shows the two areas.

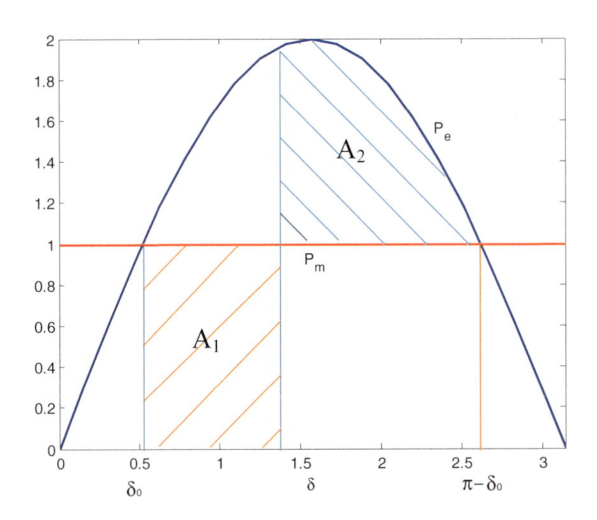

Figure 7.1: Example 1.

If the system is stable, given δ_0 and δ_1, we can find $\delta_{\max}$ using the following equation.

$$A_1 = P_m(\delta_1 - \delta_0) = A_2 = \int_{\delta_1}^{\delta_{\max}} \left(\frac{EV_\infty}{X} \sin(\delta) - P_m \right) d\delta \tag{7.11}$$

Example 2: For a SMIB system, the generator terminal bus is subject to a three-phase to ground fault at t_0. The initial rotor angle is δ_0. The fault is cleared at $t_1 = t_0 + \Delta t$. What is the maximum of Δt (or critical clearing time) to make the system stable?

The critical clearing time can be found by first making $A_1 = A_2$ and finding δ_1. Further we will express δ_1 in terms of Δt.

During the period from t_0 to t_1, $P_e = 0$. Hence the swing equation becomes

$$\ddot{\delta} = \frac{\omega_0}{2H} P_m$$

if the mechanical friction is ignored ($D_1 = 0$).

The time-domain expression of δ_1 becomes

$$\delta_1 = \delta_0 + \frac{\omega_0}{2H} P_m (\Delta t)^2$$

For a system with $H = 5$ s, $D_1 = 0$, $P_m = 1$ and $P_e = 2\sin\delta$, the initial rotor angle δ_0 (= 0.5236 rad) can be computed by considering the initial operation point as an equilibrium point.

$$\begin{aligned} 0 &= \dot{\delta} = \omega_0(\omega - 1) \quad \Rightarrow \omega = 1 \\ 0 &= \dot{\omega} = \frac{1}{2H}(P_m - P_e - D_1(\omega - 1)) = \frac{1}{2H}(1 - 2\sin\delta) \end{aligned} \tag{7.12}$$

The computed critical angle $\delta_1 = 1.3886$ rad by solving (7.11) and the corresponding critical clearing time is $\Delta t = 0.2142$ s.

Example 3: If the system is stable, we can further find the minimum rotor angle $\delta_{\min}$.

The minimum angle can be found again using the equal-area method. The starting point is the maximum rotor angle. Solving the following equation

$$0 = \int_{\delta_{\max}}^{\delta_{\min}} \left(\frac{EV_\infty}{X} \sin(\delta) - P_m \right) d\delta$$

will lead to $\delta_{\min}$. Figure 7.2 shows the two areas.

7.3.1 Time-domain simulation results

Time-domain simulation is conducted to validate the analysis results from equal-area criteria.

The following code generates the time-domain data related to the swing equation state variable vector $x = [\delta, \omega]^T$ and the generator output power

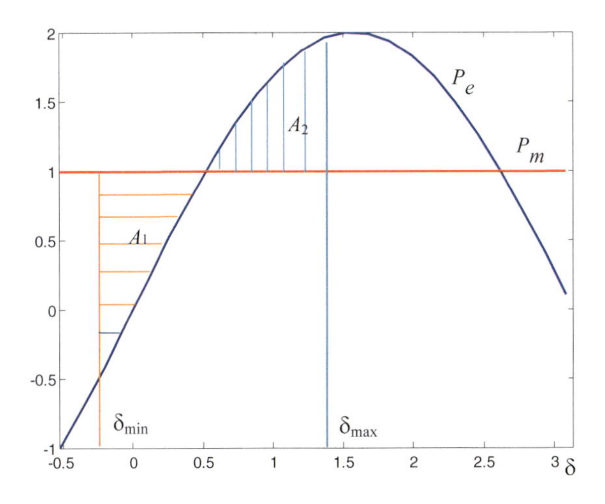

Figure 7.2: Example 3.

P_e. Note that the numerical integration method used is the trapezoidal method introduced in Chapter 2.

The input to the function is the critical clearing time tcr, time step h, and the end time T_end.

```
function [x, Pe]=SMIB_sim(tcr,h, T_end)
    omega_0= 377;
    parameter.H =5;
    parameter.D1 =0;
    delta0 = pi/6;
    E = 1; V=1; X=0.5;
    parameter.E = E;
    parameter.V = V;
    parameter.X = X;
    parameter.Pm = E*V/X*sin(delta0);
    parameter.tcr =tcr;

x(:,1) = [delta0; 1];
i=1;
for t=0:h:T_end
        i= i+1;
        [dotx, Pe(i)] = SMIB(x(:,i-1), t, parameter);
        x1 = x(:,i-1) +dotx*h;
        [dotx1, Pe(i)] = SMIB(x1, t, parameter);
        x(:,i) = x(:,i-1) +(dotx+dotx1)/2*h;
end
return
```

This function calls another function SMIB to compute $\dot{x}$ and P_e at every step. The code for SMIB is listed as follows.

```
function [x_dot, Pe] = SMIB(x, t, parameter)
t0 = 1;
x_dot = zeros(2,1);
delta = x(1); omega = x(2);
omega_0= 377;
H = parameter.H;
D1 = parameter.D1;
Pm = parameter.Pm;
tcr = parameter.tcr;
E = parameter.E;
V = parameter.V;
X = parameter.X;
if (t>t0 && t<t0+tcr)
    Pe = 0;
else
    Pe = E*V/X*sin(delta);
end
x_dot(1) = omega_0*(omega-1);
x_dot(2) = 1/(2*H)*(Pm - Pe -D1*(omega-1));
```

Results from time-domain simulation for a SMIB system subject to a three-phase ground fault at the generator bus are presented in Figures 7.3–7.5.

The following code is used to generate those figures.

```
clear; clc;
T_end = 3; h = 0.0002; T=0:h:T_end; n = length(T);
[x, Pe]= SMIB_sim(0.2140,h, T_end); x2 = x; Pe2 = Pe;
[x, Pe]= SMIB_sim(0.2142,h, T_end); x3 = x; Pe3 = Pe;
[x, Pe]= SMIB_sim(0.2145,h, T_end); x4 = x; Pe4 = Pe;

figure
plot(T, [x2(1,1:n);x3(1,1:n); x4(1,1:n)],'LineWidth',2);
ylabel('\delta'); grid on; xlabel('Time (s)'); xlim([0.9,2.2]);
figure;
plot(T, [x2(2,1:n);x3(2,1:n); x4(2,1:n)],'LineWidth',2);
ylabel('\omega'); grid on; xlabel('Time (s)');xlim([0.9,2.2]);
figure;
plot(T, [Pe2(1:n);Pe3(1:n);Pe4(1:n)],'LineWidth',2);
ylabel('P_eomega'); grid on; xlabel('Time (s)'); xlim([0.9,2.2]);
```

The simulation results validate the equal-area method analysis. It can be found that the critical clearing time is indeed 0.2142 s.

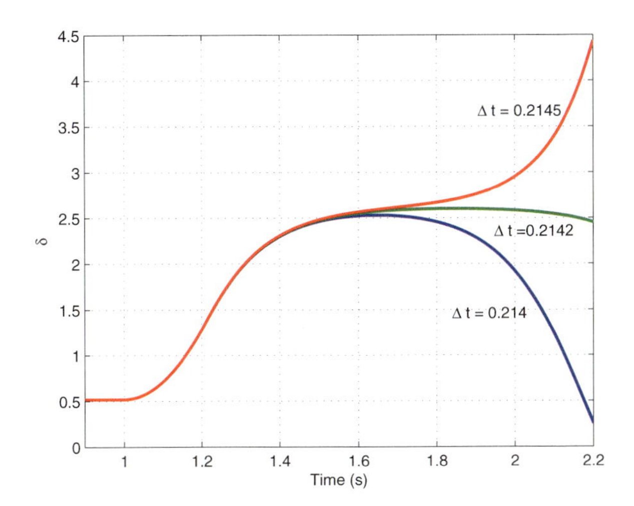

Figure 7.3: Time-domain simulation results: δ (rad).

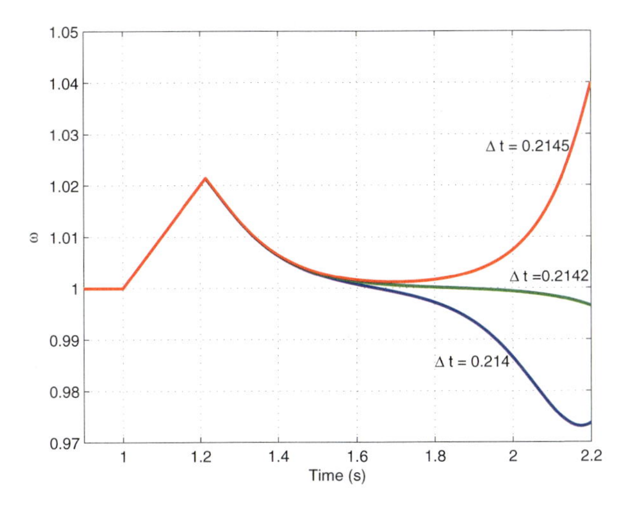

Figure 7.4: Time-domain simulation results: ω (pu).

A longer time frame is used when the clearing time is 0.2142 s. Figure 7.6 shows the simulation results for 2 seconds after the fault.

Note that at the critical clearing time, based on the equal-area method, $\delta_{\max}$ should be $\pi - \delta_0 = \frac{5}{6}\pi = 2.618$ rad. The simulation results show the maximum angle is 2.5711 rad. The error is due to numerical integration error. If we reduce the time step, this error can be reduced. For example, if

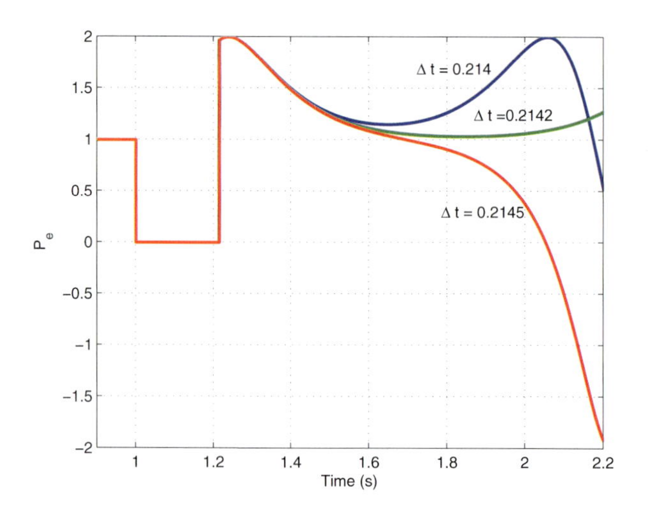

Figure 7.5: Time-domain simulation results: P_e (pu).

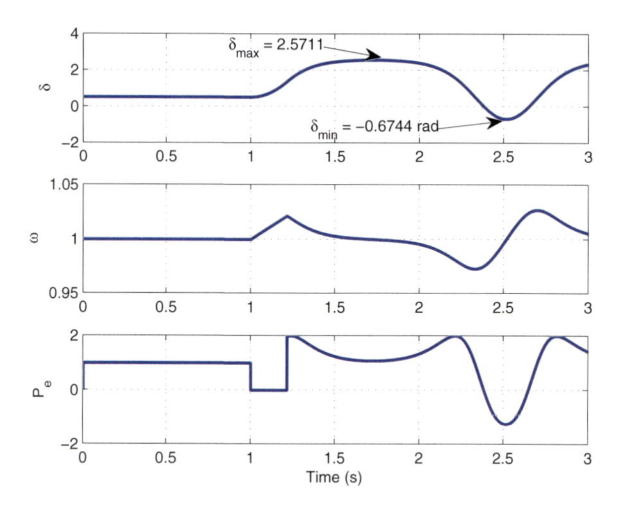

Figure 7.6: Time-domain simulation results with clearing time as 0.2142 seconds.

the time step is reduced to 10^{-5} seconds, then $\delta_{\max} = 2.595$ rad.

Exercises

1. Use the SMIB system parameter presented in the time-domain simulation example in Section 7.4. For that SMIB system, if the fault clearing time is 0.2 seconds, the initial rotor angle is 30°, $P_e = 2\sin\delta$, please compute the rotor angle at the moment when the fault is cleared δ_1, the maximum rotor angle that can be achieved δ_2, and the minimum rotor angle that can be achieved δ_{min}. Validate your analysis by examining the simulation results.

2. Modify the dynamic simulation code to include rotor flux dynamics or E'_a dynamics. Assume that $X_d = X_q = 0.3$ while the line reactance is 0.2, $X'_d = 0.15$, $E_a(0) = 1$, $T'_{d0} = 2$ seconds. Carry out the same dynamic simulation for three-phase fault and clearing.
2.1 Find the critical clearing time.
2.2 If the fault clearing time is 0.2 seconds, find from simulation the rotor angle at the moment when the fault is cleared δ_1, the maximum rotor angle that can be achieved δ_2, and the minimum rotor angle that can be achieved δ_{min}.

Compare the values you obtain in Problem 2 with those you find from simulation of Problem 1. The only difference is the inclusion of rotor flux dynamics. Comment if the modeling complexity changes the values.

Chapter 8

Small-Signal Stability

In this chapter, three example problems are presented on small-signal stability. Small-signal stability, is related to operating conditions, e.g., power transfer level. Small-signal stability influences power transfer capability of a system. In the first example, we will use small-signal stability to investigate the power transfer capability for a SMIB system with and without PSS.

In the second example, inter-area oscillations will be investigated using networked control system theory, specifically homogeneous system static output feedback stability criterion.

In the third example, we will investigate torsional interaction phenomenon between a generator's turbine and a series compensated network. Using frequency-domain models and linear system analysis tools such as the root locus method, we can show that torsional interactions occur when the series compensation degree increases.

All the above stability (instability) issues are related to operating conditions, e.g., power transfer level and compensation degree. They can all be studied using small-signal (linear) model and linear system stability analysis tools.

8.1 SMIB system stability

The SMIB system is shown in Figure 8.1. To investigate power transfer's effect on stability, the electromechanical dynamics or swing equations should be included in the model. The generator's electromagnetic dynamics will be represented by the rotor flux dynamics only. The stator electromagnetic dynamics can be ignored, which renders phasor representation of the system circuit. Automatic voltage regulator (AVR) will be included in the base model. As a comparison, a model with the addition of a PSS will also be developed.

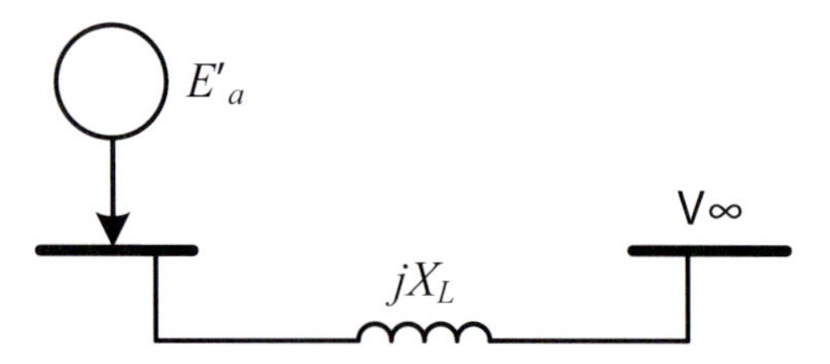

Figure 8.1: A generator connected to an infinite bus.

The block diagram of the small-signal model for the base model has been derived in Chapter 5. Here we show the block diagram of the system with AVR and PSS included in Figure 8.2. The AVR is simplified as a gain. This type of PSS is designed to amplify the derivative of the rotor speed deviation. This type of PSS appears in Bergen and Vittal (2009). Derivative of a signal is usually realized with delay. Therefore, the transfer function is $\frac{\gamma s}{1+\tau s}$, where τ is a small time constant, e.g., 0.05 s.

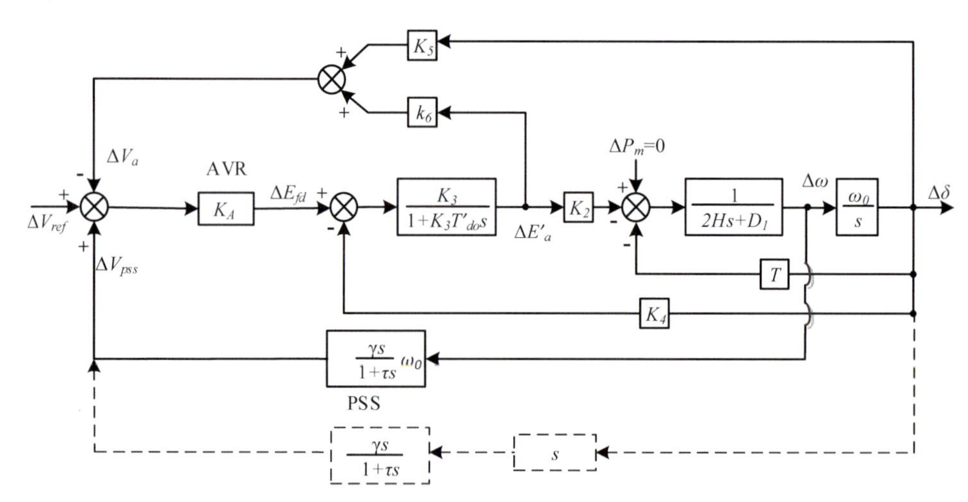

Figure 8.2: Block diagram of small-signal system with AVR and PSS included. The dotted blocks represent an alternative way of PSS representation.

The PSS block can also be represented by the dotted blocks with the rotor angle $\Delta\delta$ as input and ΔV_{pss} as the output. Such representation will simplify block aggregation for transfer function derivation.

Note that given the block diagram, it is feasible to obtain the closed-loop transfer function and examine stability by checking if the closed-loop system poles are all at LHP.

8.1.1 Computing initial state variables

The important thing to notice is that T (also notated as K_1, K_2, K_3, K_5 and K_6) are related to the system operating conditions. K_3 ($= \frac{X'_d+X_L}{X_d+X_L}$) is only related to the machine and network parameters. If the power transfer level varies, T, K_2, K_4, K_5 and K_6 will vary. The expressions of T, K_2, K_4, K_5 and K_6 are derived and listed as follows.

Power can be expressed by E'_a and δ as follows.

$$P_e = \frac{E'_a V_\infty}{\widetilde{X'_d}} \sin(\delta) + \frac{V_\infty^2}{2}\left(\frac{1}{\widetilde{X_q}} - \frac{1}{\widetilde{X'_d}}\right)\sin(2\delta) \tag{8.1}$$

Linearize the nonlinear equation by assuming small disturbances in δ and E'_a. We will have

$$\Delta P_e = \underbrace{\frac{\partial P_e}{\partial E'_a}}_{K_2} \Delta E'_a + \underbrace{\frac{\partial P_e}{\partial \delta}}_{T} \Delta\delta. \tag{8.2}$$

Therefore, T, K_2 can be found from the partial derivatives. The expressions of K_4, K_5 and K_6 have been given in Chapter 5 and are copied here.

$$T \triangleq \frac{E'_{a0} V_\infty}{\widetilde{X'_d}} \cos\delta + V_\infty^2\left(\frac{1}{\widetilde{X_q}} - \frac{1}{\widetilde{X'_d}}\right)\cos(2\delta) \tag{8.3a}$$

$$K_2 \triangleq \frac{V_\infty}{\widetilde{X'_d}} \sin\delta \tag{8.3b}$$

$$K_4 \triangleq \left(1 - \frac{\widetilde{X_d}}{\widetilde{X'_d}}\right) V_\infty \sin\delta \tag{8.3c}$$

$$K_5 \triangleq -\frac{V_\infty}{V_{a0}}\left(\frac{V_{q0} X'_d}{\widetilde{X'_d}} \sin\delta_0 + \frac{V_{d0} X_q}{\tilde{X}_q}\cos\delta_0\right) \tag{8.3d}$$

$$K_6 \triangleq \frac{V_{q0} X_L}{V_{a0} \widetilde{X'_d}} \tag{8.3e}$$

Analysis of the expressions in (8.3) shows that they are all dependent on the initial rotor angle δ. T is also dependent on the initial voltage E'_a.

K_5 and K_6 are dependent on the initial RMS value of the terminal voltage V_a as well as its dq-axis components. These values related to the terminal voltage, V_a, V_{ad}, and V_{aq}, can be determined once V_∞, E'_a and δ are known.

Note that, E'_a and δ are the state variables of system dynamics. Therefore, given the initial state variables at an equilibrium point, we should be able to find the linearized model at this operating point.

Further, usual operating conditions of a generator are defined by its power output P_e and Q_e. From P_e and Q_e, E'_a and δ need to be found.

Approach 1

Note that P_e and Q_e can both be expressed by E'_a and δ. Using the complex power computation concept, we can find P_e and Q_e.

The complex power injected to the grid can be written based on the q-axis as

$$S = V_\infty e^{-j\delta}\overline{I_a^*} = V_\infty(\cos\delta - j\sin\delta)(I_{aq} - jI_{ad}) \tag{8.4}$$

Separating the real and imaginary components, we have P_e and Q_e expressions in terms of the dq-axis currents.

$$P_e = V_\infty(\cos\delta I_{aq} - \sin\delta I_{ad}) \tag{8.5a}$$

$$Q_e = -V_\infty(\cos\delta I_{ad} + \sin\delta I_{aq}) \tag{8.5b}$$

Further, based on the phasor-diagram shown in Figure 8.3, the dq-axis components of the current can be found as (8.6).

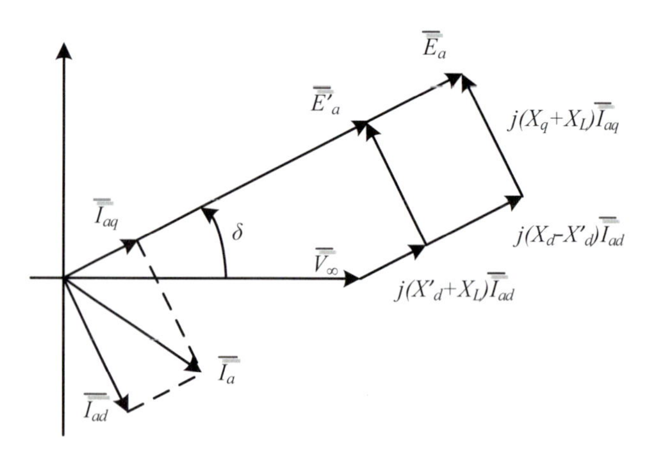

Figure 8.3: Phasor diagram.

$$I_{ad} = \frac{V_\infty \cos\delta - E'_a}{\widetilde{X}'_d} \tag{8.6a}$$

$$I_{aq} = \frac{V_\infty \sin\delta}{\widetilde{X}_q} \tag{8.6b}$$

Therefore P_e and Q_e can be expressed in terms of E'_a and δ.

$$P_e = \frac{E'_a V_\infty}{\widetilde{X}'_d}\sin\delta + \frac{V_\infty^2}{2}\left(\frac{1}{\widetilde{X}_q} - \frac{1}{\widetilde{X}'_d}\right)\sin(2\delta) \tag{8.7a}$$

$$Q_e = \frac{E'_a V_\infty}{\widetilde{X}'_d}\cos\delta - V_\infty^2\left(\frac{(\cos\delta)^2}{\widetilde{X}'_d} + \frac{(\sin\delta)^2}{\widetilde{X}_q}\right) \tag{8.7b}$$

Given P_e, Q_e, finding E'_a and δ is a problem of solving nonlinear algebraic equations. Newton–Raphson is a commonly used method for this purpose. In power systems, the Newton–Raphson method is used to solve load flow problems Bergen and Vittal (2009).

To find the solution x^* that can make $f(x) = 0$ (where $x \in \mathbb{R}^n$ and $f : \mathbb{R}^n \longrightarrow \mathbb{R}^n$), the iterative procedure of the Newton–Raphson method is written as follows.

$$x^{k+1} = x^k - \left(\frac{\partial f}{\partial x}\right)^{-1}\Bigg|_{x^k} f(x^k) \tag{8.8}$$

For this specific problem, the iterative procedure is written as follows.

$$\begin{bmatrix}\delta \\ E'_a\end{bmatrix}^{k+1} = \begin{bmatrix}\delta \\ E'_a\end{bmatrix}^{k} - J_k^{-1}\begin{bmatrix}\frac{E'^k_a V_\infty}{\widetilde{X}'_d}\sin\delta^k + \frac{V_\infty^2}{2}\left(\frac{1}{\widetilde{X}_q} - \frac{1}{\widetilde{X}'_d}\right)\sin(2\delta^k) - P_e \\ \frac{E'^k_a V_\infty}{\widetilde{X}'_d}\cos\delta^k - V_\infty^2\left(\frac{(\cos\delta^k)^2}{\widetilde{X}'_d} + \frac{(\sin\delta^k)^2}{\widetilde{X}_q}\right) - Q_e\end{bmatrix} \tag{8.9}$$

where

$$J_k = \begin{bmatrix}\frac{E'^k_a V_\infty}{\widetilde{X}'_d}\cos\delta^k + V_\infty^2\left(\frac{1}{\widetilde{X}_q} - \frac{1}{\widetilde{X}'_d}\right)\cos(2\delta^k) & \frac{V_\infty}{\widetilde{X}'_d}\sin\delta^k \\ -\frac{E'^k_a V_\infty}{\widetilde{X}'_d}\sin\delta^k + \left(\frac{1}{\widetilde{X}'_d} - \frac{1}{\widetilde{X}_q}\right)V_\infty^2\sin(2\delta^k) & \frac{V_\infty}{\widetilde{X}'_d}\cos\delta^k\end{bmatrix} \tag{8.10}$$

This algorithm is tested for an example with varying P_e and Q_e. The iterative results are shown in Figure 8.4.

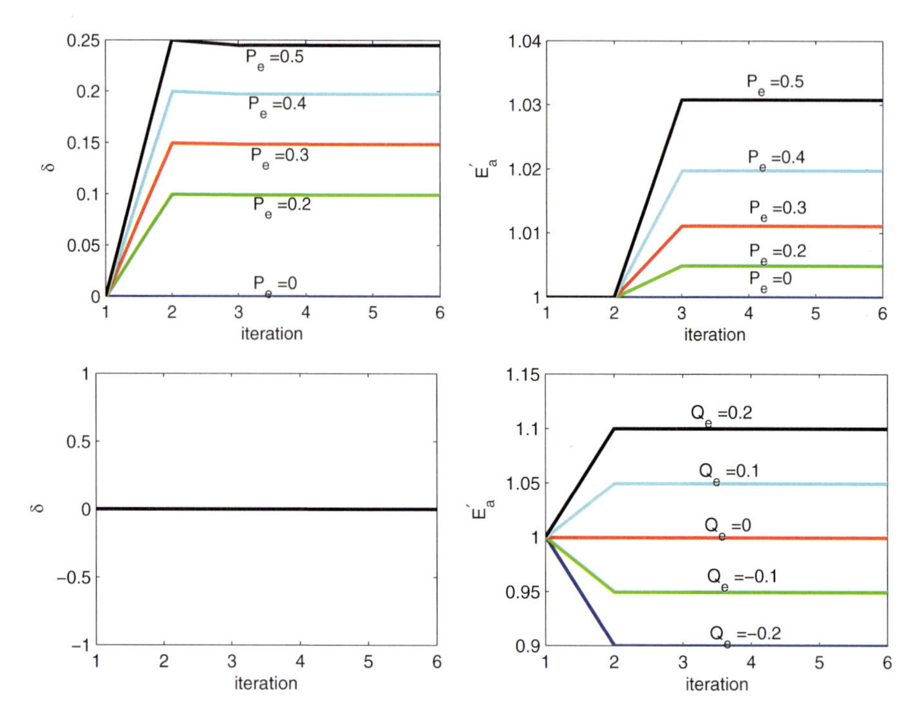

Figure 8.4: Iterative results for the Newton–Raphson algorithm. First row: $Q_e = 0$. Second row: $P_e = 0$. Other parameters: $\widetilde{X'_d} = 0.5$, $\widetilde{X_q} = 0.5$.

Below is the code to compute E'_a and δ for a given operating condition of a generator defined by P_e and Q_e. Initial values of E'_a and δ are first assumed. The Jacobian matrix is then evaluated. The iteration stops when the convergence criterion is met or the maximum iteration number is reached.

```
function [Ea1, delta] = fun_PQ2DeltaEa1(Pe, Qe, V, xd1, xq, iter_max)
% give delta and Ea_prime initial values.
x = [0; 1]; i=1;err =1;
while(i<iter_max && err > 1e-5)
    delta = x(1);
      Ea1 = x(2);
    % evalute f(x)
    fx=[Ea1*V*sin(delta)/xd1+V^2/2*(1/xq-1/xd1)*sin(2*delta)-Pe;
        Ea1*V*cos(delta)/xd1-V^2*((cos(delta))^2/xd1+(sin(delta))^2/xq)-Qe];
    % error
    err = max(abs(fx));
    data(:,i) = [x; fx; err];
    % evalute J(x)
    J = zeros(2,2);
```

```
    J(1,1) = Ea1*V*cos(delta)/xd1+V^2*(1/xq-1/xd1)*cos(2*delta);
    J(1,2) = V*sin(delta)/xd1;
    J(2,1) = -Ea1*V*sin(delta)/xd1+V^2*(1/xd1-1/xq)*sin(2*delta);
    J(2,2) = V*cos(delta)/xd1;
    % update x
    dx =  - inv(J)*fx;
    x = x + dx;
    i= i+1;
end
```

Approach 2

Approach 1 requires an iterative procedure to solve a nonlinear algebraic equation. To initialize a generator's state variables, there is a direct approach. Given terminal voltage and terminal current, applying phasor diagram can lead to E'_a and δ. Iteration is not needed. If the terminal voltage, real power and reactive power are given, then the terminal current is also given.

For the SMIB system, if the complex power to the grid is given, then the terminal current is given as

$$\overline{I}_a = \frac{P_e - jQ_e}{V_\infty} = I_a e^{j\theta_a}. \tag{8.11}$$

We can first find the rotor's position relative to the infinity bus voltage's space vector by computing an intermediate voltage phasor V'.

$$\overline{V}' = V_\infty e^{j0} + j(X_L + X_q)\overline{I}_a = V' e^{j\delta}. \tag{8.12}$$

The angle of $\overline{V}'$ is the rotor angle δ. The current is then decomposed into dq-axis components:

$$\begin{aligned} I_{aq} &= I_a \cos(\theta_a - \delta) \\ I_{sd} &= I_a \sin(\theta_a - \delta) \end{aligned} \tag{8.13}$$

Finally, we can find E'_a from the phasor diagram in Figure 8.3:

$$E'_a = V_\infty \cos\delta - (X'_d + X_L) I_{ad} \tag{8.14}$$

8.1.2 Computation of the linearized model parameters

With the state variables E'_a and δ at the initial operating condition known, we then compute all parameters related to the operating condition and the linear model. Below is the code to compute T, K_2, K_4, K_5, K_6 for a given operating condition of a generator defined by E'_a and δ.

```
function K = compute_K_SMIB(Ea1, delta, Xd, Xd1, Xq, XL)
Vinf=1;
Xdz=Xd+XL; Xqz=Xq+XL; Xdprimez=Xd1+XL;
Iad=(Vinf*cos(delta)-Ea1)./Xdprimez;
Iaq=(Vinf*sin(delta))./Xqz;
Vaq=Vinf*cos(delta)-XL*Iad Vad=XL*Iaq-Vinf*sin(delta);
Va=(Vad.^2+Vaq.^2).^(1/2);

T=Ea1*Vinf/Xdprimez.*cos(delta)+Vinf^2*(1/Xqz-1/Xdprimez)*cos(2*delta);
K2=Vinf/Xdprimez*sin(delta);
K3=Xdprimez/Xdz;
K4=(1/K3-1)*Vinf*sin(delta);
K5=-Vinf./Va.*(Vaq*Xd1/Xdprimez.*sin(delta)+Vad*Xq/Xqz.*cos(delta));
K6=Vaq*XL./(Va*Xdprimez);
K = [T;K2; K4; K5; K6];
```

Using the above codes, for a given set of P_e and Q_e, the corresponding coefficients (T and Ks) can be found. Figure 8.5 presents the T and Ks for a varying P_e at three scenarios: leading power factor, unity power factor, and lagging power factor. It is shown from Figure 8.5 that K_5 changes sign when the active power demand from the generator becomes heavy.

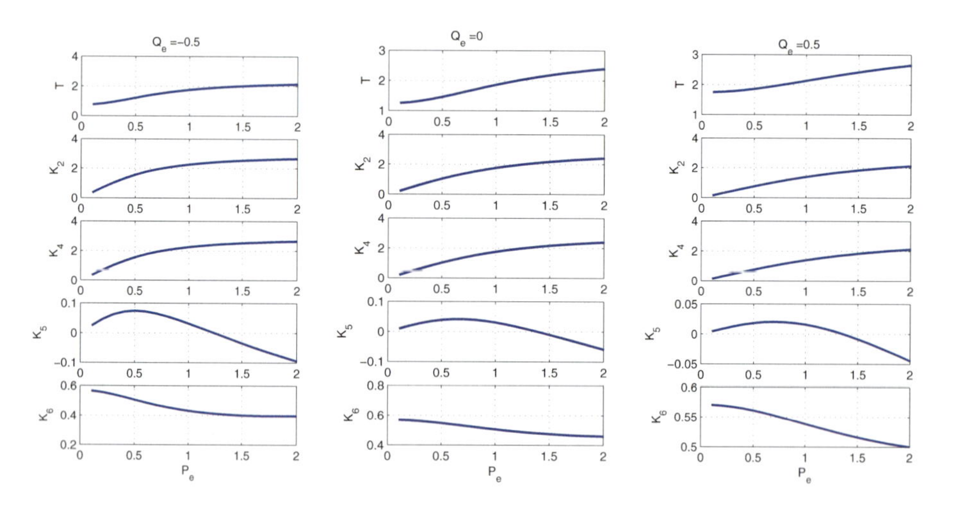

Figure 8.5: T and Ks for a varying P_e. Left: $Q_e = -0.5$; Middle: $Q_e = 0$; Right: $Q_e = 0.5$.

8.1.3 Linearized model without PSS

In this subsection, PSS is not considered. Stability analysis will be carried for a system shown in Figure 8.6. ΔP_m is the input, $\Delta\delta$ is the output, and ΔV_{ref} is normally zero.

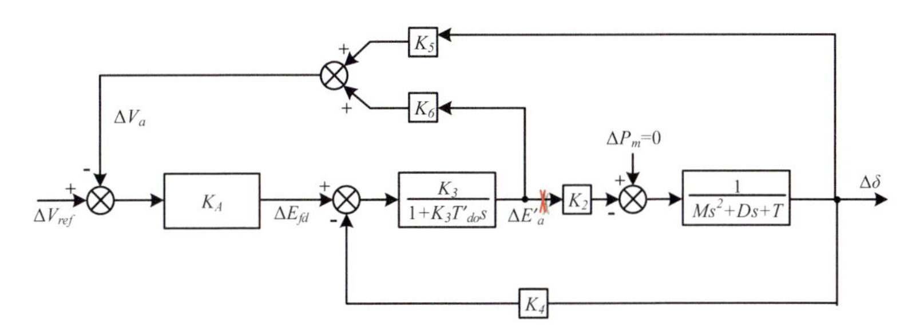

Figure 8.6: Block diagram of a linearized system without PSS.

There are two assumptions which are made to simplify the model. It is assumed that the voltage control system is very fast, so the time constant of the control loop is zero; the exciter is also assumed to be a pure gain 1.

Figure 8.6 will be aggregated into a single loop system. Then the closed-loop transfer function will be derived. First, the loop will be decoupled at the point before the block of K_2. For the open-loop system, we will compute the transfer function from $\Delta\delta$ to $\Delta E'_a$. Then we close the loop and compute the closed-loop transfer function by considering the feedforward block from $\Delta E'_a$ to $\Delta\delta$. The feedforward transfer function is notated as $-k_2 T_{EM}$ where $T_{EM} = 1/(Ms^2 + Ds + T)$.

$$\begin{aligned}
\Delta E'_a &= \frac{K_3}{T'_{do}K_3 s + 1}[-K_4\Delta\delta - K_A(K_5\Delta\delta + K_6\Delta E'_a)] \\
\frac{\Delta E'_a}{\Delta\delta} &= \frac{-K_3(K_4 + K_A K_5)}{T'_{do}K_3 s + K_3 K_A K_6 + 1} \\
&= -\frac{1}{K_2}\frac{a}{bs + c}
\end{aligned} \tag{8.15}$$

where $a = K_2K_3(K_4 + K_AK_5)$, $b = K_3T_{do}$, $c = K_3K_AK_6 + 1$.

The block diagram in Figure 8.6 becomes Figure 8.7. Root loci of the open-loop transfer function $\frac{a}{bs+c}\frac{-1}{Ms^2+Ds+T}$ are plotted for two scenarios. When $K_4 + K_AK_5 > 0$ or $a > 0$, the closed-loop system is in fact a positive-feedback system since the steady-state gain of the open-loop system is $\frac{-a}{T} < 0$. When the system has a heavy power transfer, K_5 will be less than 0 and $a < 0$. In this scenario, the system is a negative-feedback system. It can be seen that when $a < 0$, the system may suffer oscillations since the two complex conjugate poles related to the electromechanical dynamics move to the RHP.

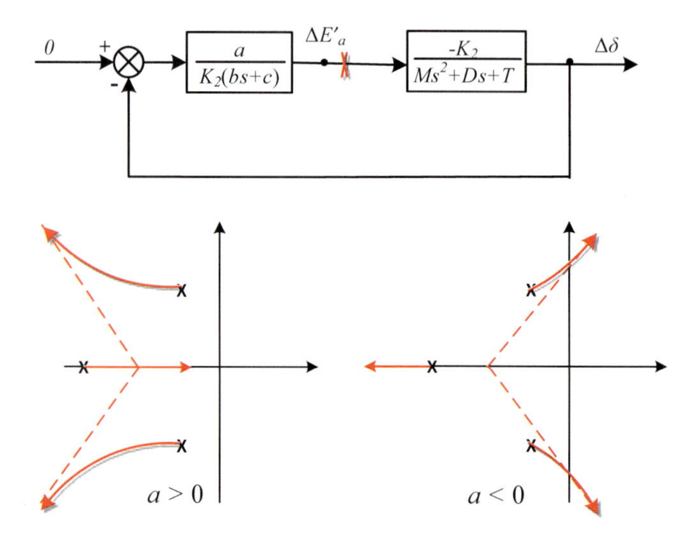

Figure 8.7: Single-loop block diagram of linearized system without PSS and the open-loop gain's root loci.

The closed-loop transfer function is derived as follows.

$$\begin{aligned}
T_{cl}(s) &= \frac{-K_2 \frac{1}{Ms^2++Ds+T}}{1 + K_2 \frac{1}{Ms^2++Ds+T}\left(-\frac{1}{K_2}\frac{a}{bs+c}\right)} \\
&= \frac{-K_2(bs+c)}{(Ms^2++Ds+T)(bs+c)-a} \\
&= \frac{-K_2(bs+c)}{Mbs^3+(Db+Mc)s^2+(Tb+Dc)s+Tc-a}
\end{aligned} \tag{8.16}$$

Based on $T_{cl}(s)$, there are three poles and one zero. Stability of the closed-loop system is determined by the poles of $T_{cl}(s)$ or the roots of the denominator polynomial. This polynomial is also called a characteristic equation. If all the poles are located in LHP, then the system is stable. Therefore, stability can be checked by computing the roots of the denominator polynomial. In addition, we can rely on the root locus method by plotting root loci of the open-loop gain.

We can also directly check the coefficients of the denominator polynomial to determine stability, applying Routh–Hurwitz stability criterion.

The Routh–Hurwitz criterion is a sufficient and necessary condition for stability. Given a characteristic equation

$$a_n s^n + a_{n-1} s^{n-1} + \cdots + a_1 s + a_0 = 0, \tag{8.17}$$

an array will be arranged as follows Dorf and Bishop (1998):

$$\begin{array}{c|cccc} s^n & a_n & a_{n-2} & a_{n-4} \cdots \\ s^{n-1} & a_{n-1} & a_{n-3} & a_{n-5} \cdots \\ s^{n-2} & b_{n-1} & b_{n-3} & b_{n-5} \cdots \\ s^{n-3} & c_{n-1} & c_{n-3} & c_{n-5} \cdots \\ \cdot & \cdot & \cdot & \cdot \\ \cdot & \cdot & \cdot & \cdot \\ s^0 & h_{n-1} & & \end{array}$$

where

$$b_{n-1} = \frac{-1}{a_{n-1}} \begin{vmatrix} a_n & a_{n-2} \\ a_{n-1} & a_{n-3} \end{vmatrix}$$

$$b_{n-3} = \frac{-1}{a_{n-1}} \begin{vmatrix} a_n & a_{n-4} \\ a_{n-1} & a_{n-5} \end{vmatrix}$$

$$c_{n-1} = \frac{-1}{b_{n-1}} \begin{vmatrix} a_n & a_{n-3} \\ b_{n-1} & b_{n-3} \end{vmatrix}$$

and so on.

The Routh–Hurwitz criterion: The number of roots of the characteristic equation with positive real parts is equal to the number of the changes in sign of the first column of the Routh array. If the first column coefficients are all greater than 0 or if the first column coefficients are all less than 0, the system has 0 number of roots with positive real parts and the system is stable.

The Routh array for $Mbs^3 + (Db + Mc)s^2 + (Tb + Dc)s + Tc - a = 0$ is as follows.

$$\begin{array}{c|ccc} s^3 & Mb & Tb + Dc & 0 \\ s^2 & Db + Mc & Tc - a & 0 \\ s^1 & Tb + Dc - \frac{(Mb)(Tc-a)}{Db+Mc} & 0 & \\ s^0 & Tc - a & & \end{array}$$

We can see the stability is determined by the first column coefficients. Those coefficients, Mb, $Db + Mc$, $Tb + Dc - \frac{(Mb)(Tc-a)}{Db+Mc}$, and $Tc - a$ should all be greater than zero to guarantee stability. Note that K_5 can be negative when the power demand is heavy and a is dependent on K_5. It can be seen that a heavy power demand may cause the coefficient $Tb + Dc - \frac{(Mb)(Tc-a)}{Db+Mc}$ to be negative. Thus the system will be unstable. Further, the sign of the first column coefficients will change twice: from $DB + MC > 0$ to $Tb + Dc - \frac{(Mb)(Tc-a)}{Db+Mc} < 0$, then to $Tc - a > 0$. The system will have two poles with positive real part.

8.1.4 Linearized model with PSS

A PSS is added to stabilize the system. Using a similar technique to derive the closed-loop transfer function of system, first, we break the loop at the point before the block at K_2. The feedback transfer function from $\Delta\delta$ to $\Delta E'_a$ can be found.

$$\begin{aligned}
\Delta E'_a &= \frac{K_3}{T'_{do}K_3 s + 1}\left[K_A\left(\frac{\gamma s^2}{\tau s + 1}\Delta\delta - K_5\Delta\delta - K_6\Delta E'_a\right)\right] \\
&\quad - \frac{K_3}{T'_{do}K_3 s + 1}K_4\Delta\delta \\
\Rightarrow \frac{\Delta E'_a}{\Delta\delta} &= \frac{K_3 k_e \gamma s^2 - \tau K_3(K_5K_A + K_4)s - K_3(K_5K_A + K_4)}{\tau T'_{do}K_3 s^2 + (\tau K_3K_6K_A + \tau + T'_{do}K_3)s + K_3K_6K_A + 1} \\
\Rightarrow \frac{\Delta E'_a}{\Delta\delta} &= \frac{1}{K_2}\frac{fs^2 - \tau as - a}{\tau bs^2 + es + c} \qquad (8.18)
\end{aligned}$$

where γ and τ are the coefficients of PSS, $e = \tau k_3k_6k_e + \tau + T'_{do}k_3$ and $f = k_3k_e\gamma k_2$. The closed-loop transfer function is written as:

$$\begin{aligned}
T_{cl}(s) &= \frac{-K_2\frac{1}{Ms^2+Ds+T}}{1 + \frac{1}{Ms^2+Ds+T}\frac{fs^2-\tau as-a}{\tau bs^2+es+c}} \\
&= \frac{\tau bs^2 + es + c}{(Ms^2 + Ds + T)(\tau bs^2 + es + c) + fs^2 - \tau as - a} \qquad (8.19)
\end{aligned}$$

Stability will be determined by the denominator

$$M\tau bs^4 + (Me + D\tau b)s^3 + (Mc + De + T\tau b + f)s^2 + (Dc + Te - \tau a)s + (Tc - a).$$

The Routh array is shown as follows.

$$\begin{array}{c|cccc}
s^4 & M\tau bs & Mc + De + T\tau b + f & Tc - a & 0 \\
s^3 & Me + D\tau b & Dc + Te - \tau a & 0 & 0 \\
s^2 & K_{r2} & Tc - a & 0 & \\
s^1 & K_{r1} & 0 & & \\
s^0 & Tc - a & & &
\end{array}$$

where

$$K_{r2} = Mc + De + T\tau b + f - \frac{(M\tau b)(Dc + Te - \tau a)}{Me + D\tau b},$$

$$K_{r1} = Dc + Te - \tau a - \frac{(Me + D\tau b)(Tc - a)}{K_{r2}}.$$

Based on the calculation, K_{r2} and K_{r1} are always larger than 0.

This shows that with PSS, the system will be stable.

8.2 Inter-area oscillations

Inter-area oscillations are defined as a group of generators swinging against other group(s) of generators. This type of oscillations is the root cause of 1996's blackout in the western system Kosterev et al. (1999). Small-signal stability analysis is a standard approach for inter-area oscillations.

The conventional approach is to build the state-space model of the entire system consisting of multiple generators Rogers (2012). Eigenvalue and participation factor analysis are then carried out to identify the related states to an oscillation mode Rogers (2012).

8.2.1 Consensus control

Recent advances in consensus control of multi-agents over a network provide a different approach to shed insights. In this section, results from the author's paper Fan (2017) are summarized.

The main stability criterion used in Fan (2017) is applicable for homogeneous system consensus control through static output feedback Hengster– Movric et al. (2015). The homogeneous subsystems are defined as follows.

$$\dot{x}_i = Ax_i + Bu_i \tag{8.20}$$

$$y_i = Cx_i \tag{8.21}$$

where $i = 1, \cdots, n$.

The entire system consists of n subsystems. Each subsystem is identical. Input of each subsystem u_i is expressed as follows.

$$u_i = \sum_{j \neq i}^{n} a_{ij}(y_j - y_i) \tag{8.22}$$

where $a_{ij} > 0$.

In the form of vectors, we have

$$u = Ly \tag{8.23}$$

where L is called a Laplacian matrix .

$$L_{ij} = \begin{cases} -a_{ij}, & i \neq j, \quad i,j \text{ are connected through a link} \\ 0, & i \neq j, \quad i,j \text{ are not connected} \end{cases} \tag{8.24}$$

$$L_{ii} = \sum_j a_{ij}, \quad i \neq j \tag{8.25}$$

The entire system can be written as

$$\dot{x} = (I \otimes A + L \otimes BC)x. \tag{8.26}$$

where $\otimes$ notates Kroneck product.

Research in Hengster– Movric et al. (2015) gives the stability criterion for the above system. The system $(I \otimes A + L \otimes BC)$ is Hurwitz if and only if all the matrices $A + \lambda_i BC$ are Hurwitz, where λ_i are the eigenvalues of L.

This stability criterion will be adopted for power system inter-area oscillation analysis. To apply the analysis technique, the power system will be converted to a networked control problem with homogenous systems and static output feedback.

Example The above stability criterion will be explained using a three-battery control example. Assume that each battery's power order is controlled through an integral control. We will design the input to the integral control. Also assume that the power control dynamics is fast so that power order and the power measurement are equivalent. The objective of the consensus control is for the three batteries achieve the same energy level and power output level.

First, the system dynamic model is introduced.

$$\begin{cases} \dot{E}_1 = -P_1 \\ \dot{P}_1 = U_1 \\ \dot{E}_2 = -P_2 \\ \dot{P}_2 = U_2 \\ \dot{E}_3 = -P_3 \\ \dot{P}_3 = U_3 \end{cases} \tag{8.27}$$

where E_i notates the energy level in ith battery, P_i notates the power level in ith battery and U_i is the integral control's input. In addition, we should keep the total power output constant at any time to meet the load demand.

Therefore, the control objectives are:

$$\frac{d(P_1 + P_2 + P_3)}{dt} = \Sigma U_i = 0, \text{at any time} \tag{8.28}$$

$$E_1 = E_2 = E_3, P_1 = P_2 = P_3, \text{ at final steady-state} \tag{8.29}$$

Assume that the communication graph of the batteries is as Figure 8.8.

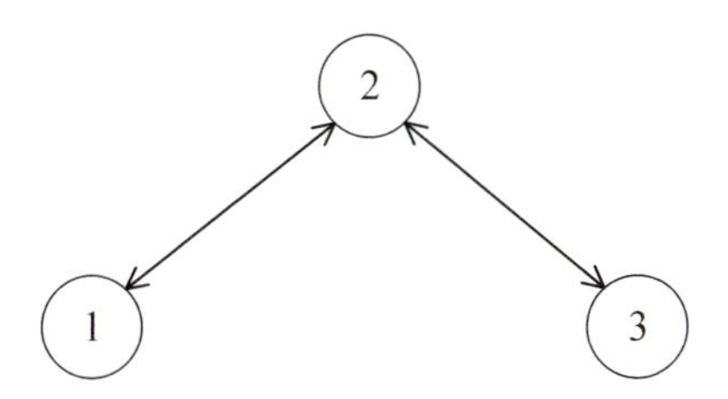

Figure 8.8: Communication topology for the three batteries.

The graph Laplacian is written as follows.

$$L = \begin{bmatrix} 1 & -1 & 0 \\ -1 & 2 & -1 \\ 0 & -1 & 1 \end{bmatrix} \tag{8.30}$$

Note that the input of the integral control will be zero when $t \to \infty$. Therefore, we will use the difference of energy levels and the difference of power levels as the input to the integral control. Further, based on the communication topology, the inputs are chosen as follows.

$$\begin{aligned} U_1 &= K_{1E}(E_1 - E_2) + K_{1P}(P_1 - P_2) \\ U_2 &= K_{2E}(E_2 - E_1) + K_{2P}(P_2 - P_1) + K'_{2E}(E_2 - E_3) + K'_{2P}(P_2 - P_3) \\ U_3 &= K_{3E}(E_3 - E_2) + K_{3P}(P_3 - P_2) \end{aligned} \tag{8.31}$$

$U_1 + U_2 + U_3 = 0$ should be satisfied at any time. We can make $K_{1E} = K_{2E}$, $K_{1P} = K_{2P}$, $K'_{2E} = K_{3E}$, and $K'_{2P} = K_{3P}$. We will introduce K_E, K_P, K'_E and K'_P. Then:

$$\begin{bmatrix} U_1 \\ U_2 \\ U_3 \end{bmatrix} = \begin{bmatrix} K_E & K_P & 0 & 0 \\ K_E & K_P & K'_E & K'_P \\ 0 & 0 & K'_E & K'_P \end{bmatrix} \begin{bmatrix} E_1 - E_2 \\ P_1 - P_2 \\ E_3 - E_2 \\ P_3 - P_2 \end{bmatrix} \tag{8.32}$$

If we let $K'_E = K_E$ and $K'_P = K_P$, then

$$\begin{bmatrix} U_1 \\ U_2 \\ U_3 \end{bmatrix} = \begin{bmatrix} K_E & K_P & 0 & 0 \\ K_E & K_P & K_E & K_P \\ 0 & 0 & K_E & K_P \end{bmatrix} \begin{bmatrix} E_1 - E_2 \\ P_1 - P_2 \\ E_3 - E_2 \\ P_3 - P_2 \end{bmatrix}$$

$$= \begin{bmatrix} K_E & 0 \\ K_E & K_E \\ 0 & K_E \end{bmatrix} \begin{bmatrix} E_1 - E_2 \\ E_3 - E_2 \end{bmatrix} + \begin{bmatrix} K_P & 0 \\ K_P & K_P \\ 0 & K_P \end{bmatrix} \begin{bmatrix} P_1 - P_2 \\ P_3 - P_2 \end{bmatrix} \tag{8.33}$$

$$\Longrightarrow U = K_E \cdot L \cdot \begin{bmatrix} E_1 \\ E_2 \\ E_3 \end{bmatrix} + K_P \cdot L \cdot \begin{bmatrix} P_1 \\ P_2 \\ P_3 \end{bmatrix}$$

$$= \begin{bmatrix} K_E & K_P & -K_E & -K_P & 0 & 0 \\ -K_E & -K_P & 2K_E & 2K_P & -K_E & -K_P \\ 0 & 0 & -K_E & -K_P & K_E & K_P \end{bmatrix} \begin{bmatrix} E_1 \\ P_1 \\ E_2 \\ P_2 \\ E_3 \\ P_3 \end{bmatrix} \tag{8.34}$$

$$= (L \otimes \underbrace{\begin{bmatrix} K_E & K_P \end{bmatrix}}_{K})x \tag{8.35}$$

where $x = [E, P]^T$. $\otimes$ stands for Kroneck product.

A Kroneck product is defined as the following:

$$A \otimes B = \begin{bmatrix} A_{11}B & A_{12}B & \cdots & A_{1n}B \\ \vdots & \vdots & \vdots & \vdots \\ A_{m1}B & A_{m2}B & \cdots & A_{mn}B \end{bmatrix} \tag{8.36}$$

For each agent, the state-space model is as follows.

$$\dot{x}_i = Ax_i + BU_i,$$
$$A = \begin{bmatrix} 0 & -1 \\ 0 & 0 \end{bmatrix}, \quad B = \begin{bmatrix} 0 \\ 1 \end{bmatrix} \tag{8.37}$$

Then the entire interconnected system has the following state-space model.

$$\dot{x} = \begin{bmatrix} A & 0 & 0 \\ 0 & A & 0 \\ 0 & 0 & A \end{bmatrix} \begin{bmatrix} E_1 \\ P_1 \\ E_2 \\ P_2 \\ E_3 \\ P_3 \end{bmatrix} + \begin{bmatrix} B & 0 & 0 \\ 0 & B & 0 \\ 0 & 0 & B \end{bmatrix} \begin{bmatrix} U_1 \\ U_2 \\ U_3 \end{bmatrix} \tag{8.38}$$

$$= \left(\begin{bmatrix} A & 0 & 0 \\ 0 & A & 0 \\ 0 & 0 & A \end{bmatrix} + \begin{bmatrix} BK & -BK & 0 \\ -BK & 2BK & -BK \\ 0 & -BK & BK \end{bmatrix} \right) x \tag{8.39}$$

$$= \begin{bmatrix} A+BK & -BK & 0 \\ -BK & A+2BK & -BK \\ 0 & -BK & A+BK \end{bmatrix} x \tag{8.40}$$

$$= (I_n \otimes A + L \otimes BK)X \tag{8.41}$$

The dynamics of the above system (with homogeneous agents) has been analyzed in Fax and Murray (2004). In short, the dynamics (or eigenvalues) of the above system can be found from the following systems:

$$A + \lambda_i BK \tag{8.42}$$

where λ_i is the eigenvalue of the Laplacian matrix L. In our case,

$$\lambda = \{0, 1, 3\}. \tag{8.43}$$

The eigenvalues of the system matrix in (8.41) will be computed and compared with the eigenvalues computed from three matrices in (8.42). Table 8.1 shows that the eigenvalues computed from both ways are exactly the same. Using (8.42) has the advantage in computing, where small-size matrices are dealt with.

8.2.2 Power system viewed as a networked control problem

Consider a system with n generators. For every generator in the power system, a classic model is assumed. The dynamics of each subsystem is expressed in a state-space model, where δ notates rotor angle in rad, ω

Table 8.1: Eigenvalues comparison

$I_n \otimes A + L \otimes BK$	$A + \lambda_i BK$	λ_i
$0.7913 + j0.0000$	$0.7913 + j0.0000$	3
$-3.7913 + j0.0000$	$-3.7913 + j0.0000$	3
$-1.6180 + j0.0000$	$-1.6180 + j0.0000$	1
$0.6180 + j0.0000$	$0.6180 + j0.0000$	1
$0.0000 + j0.0000$	$0.0000 + j0.0000$	0
$0.0000 - j0.0000$	$0.0000 + j0.0000$	0

notates speed in pu, and P_D notates area load in pu.

$$\underbrace{\begin{bmatrix} \dot{\Delta\delta_i} \\ \dot{\Delta\omega_i} \end{bmatrix}}_{\dot{x}_i} = \underbrace{\begin{bmatrix} 0 & \omega_0 \\ 0 & -\frac{D_i}{H_i} \end{bmatrix}}_{A_i} \underbrace{\begin{bmatrix} \Delta\delta_i \\ \Delta\omega_i \end{bmatrix}}_{x_i} + \underbrace{\begin{bmatrix} 0 \\ \frac{-1}{2H_i} \end{bmatrix}}_{B_i} \underbrace{\Delta P_{\text{tie,i}}}_{u_i} + \begin{bmatrix} 0 \\ \frac{-1}{2H_i} \end{bmatrix} \underbrace{\Delta P_{D,i}}_{d_i}$$

$$\underbrace{\Delta\delta_i}_{y_i} = \underbrace{\begin{bmatrix} 1 & 0 \end{bmatrix}}_{C} \begin{bmatrix} \Delta\delta_i \\ \Delta\omega_i \end{bmatrix} \tag{8.44}$$

If we treat the total tie-line flow $P_{\text{tie,i}}$ as the system's input u_i, and the rotor angle $\Delta\delta_i$ as the output y_i, then the entire system can be viewed as a consensus or synchronizing control over a network. The input u_i has the following structure.

$$u_i = \sum_j \Delta P_{ij} = \sum_{j \neq i}^{n} T_{ij}(\Delta\delta_i - \Delta\delta_j) = \sum_{j \neq i}^{n} T_{ij}(y_i - y_j)$$

where P_{ij} is the tie-line flow on the line between bus i and bus j and $T_{ij} = \frac{\partial P_{ij}}{\partial \delta_{ij}}$.

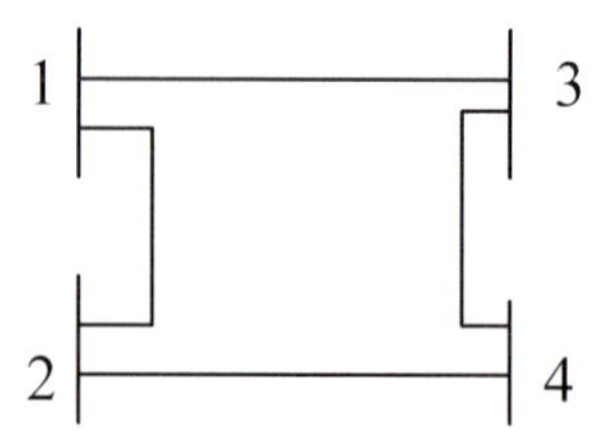

Figure 8.9: A four-bus network.

For a network shown in Figure 8.9, we will have the input vector u as the following.

$$\begin{bmatrix} u_1 \\ u_2 \\ u_3 \\ u_4 \end{bmatrix} = \underbrace{\begin{bmatrix} T_{12}+T_{13} & -T_{12} & -T_{13} & 0 \\ -T_{12} & T_{12}+T_{24} & 0 & -T_{24} \\ -T_{13} & 0 & T_{13}+T_{34} & -T_{34} \\ 0 & -T_{24} & -T_{34} & T_{24}+T_{34} \end{bmatrix}}_{L} \begin{bmatrix} y_1 \\ y_2 \\ y_3 \\ y_4 \end{bmatrix} \tag{8.45}$$

Note that L is a weighted graph Laplacian matrix. If every generator has the same parameters, then we have homogeneous systems and $A_i = A, B_i = B$. Therefore

$$\dot{x} = (I \otimes A + L \otimes BC)x + (I \otimes B)d \tag{8.46}$$

8.2.3 Case study

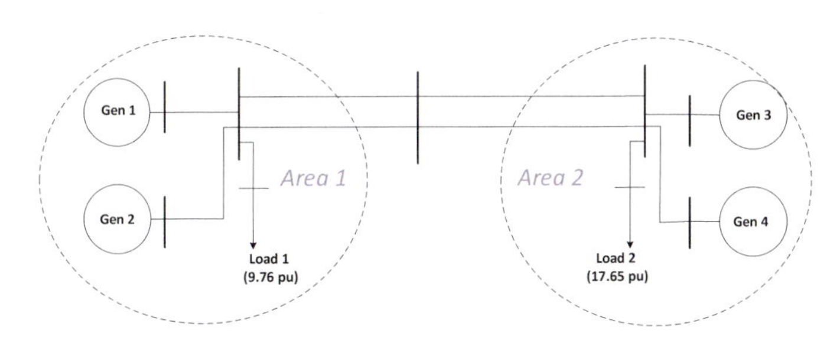

Figure 8.10: Two-area four-machine test system.

The two-area four-machine power system used for inter-area oscillations is shown in Figure 8.10. If the tie-lines are very long, the power grid connection of the system can be converted to the connection in Figure 8.9. Therefore, we can adopt the graph Laplacian matrix L in (8.45) for stability analysis.

If we assume that the system symmetric and $T_{13} = T_{24}$, $T_{12} = T_{34}$, then the eigenvalues of L are:

$$\begin{bmatrix} \lambda_1 & \lambda_2 & \lambda_3 & \lambda_4 \end{bmatrix} = \begin{bmatrix} 0 & 2T_{13} & 2T_{12} & 2(T_{13}+T_{12}) \end{bmatrix} \tag{8.47}$$

It can be seen that λ_2, the second smallest eigenvalue of the Laplacian matrix is related to T_{13}. For simplicity of analysis, assume that the line is lossless. Then

$$P_{13} = \frac{E_1 E_3}{X_{13}} \sin(\delta_1 - \delta_3) \tag{8.48}$$

$$T_{13} = \frac{E_1 E_3}{X_{13}} \cos(\delta_1 - \delta_3) \tag{8.49}$$

A longer line corresponds to a greater line reactance X_{13}, in turn a smaller λ_2. A heavier power transfer corresponds to a greater angle difference $\delta_1 - \delta_3$ and also a smaller T_{13} or λ_2.

On the other hand, the other two eigenvalues λ_3 and λ_4 are dominant by T_{12} as T_{12} is much greater than T_{13} due to the close connection between Gen 1 and Gen 2.

The system eigenvalues are determined by the following matrices' eigenvalues:

$$A - \lambda_i BC = \begin{bmatrix} 0 & \omega_0 \\ \frac{-\lambda_i}{2H} & \frac{-D}{2H} \end{bmatrix} \tag{8.50}$$

In turn, the system eigenvalues are also determined by the following polynomials.

$$s^2 + \frac{D}{2H}s + \frac{\lambda_i \omega_0}{2H} = 0 \tag{8.51}$$

$$s \approx -\frac{\sqrt{2}}{4}\frac{1}{\sqrt{\lambda_i H \omega_0}} \pm j\sqrt{\frac{\lambda_i \omega_0}{2H}} \tag{8.52}$$

Observing (8.52), we can see that for a small λ_i, the oscillation mode's frequency is low, while for a large λ_i, the oscillation mode's frequency will be large. This indicates that the low frequency inter-area oscillation corresponds to the oscillation mode associated to λ_2, while the local oscillation modes are associated to the other Laplacian matrix eigenvalues.

Table 8.2 lists the eigenvalues computed by Power System Toolbox using the conventional overall system analysis approach versus the eigenvalues computed using small-scale matrices in (8.50) at a typical operation case (case 1) and the same operation case with one of the tie-lines tripped (case 2). The comparison shows that the proposed small-scale matrix eigenvalue calculation can capture the system dynamics adequately.

8.3 Subsynchronous resonances

In this section, subsynchronous resonances (SSR) are presented. SSRs are related to series compensated electric networks. The type of resonances is due to the interaction of the mechanical oscillation mode of a synchronous generator's rotor shaft and the LC resonance mode.

Torsional interaction of synchronous generators can result in rotor shaft fracture. Such incidents happened in 1970s in U.S. at the Mohave power plant of the Southern California Edison Company Walker et al. (1975). The

Table 8.2: Eigenvalues computed by PST and the proposed method for two cases

case 1			case 2		
PST	proposed	$\lambda_i(L)$	PST	proposed	$\lambda_i(L)$
-0.0111	-0.0000	0.0000	$.0126$	$.0000$	-0.00
0.0111	-0.0077		$-.0126$	$-.0077$	
$-.00 \pm j3.53$	$-.0038 \pm j3.60$	4.0221	$-.00 \pm j2.04$	$-.0038 \pm j2.15$	1.43
$.00 \pm j7.51$	$-.0038 \pm j7.50$	17.48	$.00 \pm j7.49$	$-.0038 \pm j7.40$	16.99
$-.00 \pm j7.57$	$-.0038 \pm j8.55$	22.68	$-.00 \pm j7.52$	$-.0038 \pm j8.14$	20.59

resonances were mitigated by reducing the compensation level of the electric network and installing a torsional relay.

SSRs can be studied using linear system analysis. In the literature, torsional interactions were studies using state-space models and eigenvalue, participation factor analysis (see Chapter 15 in Kundur et al. (1994)). In this text, frequency-domain based analysis is adopted for its simplicity and the ability of shedding insights. First we build transfer functions block by block. Then we study the structure of the system using those blocks. And finally we can adopt classic control analysis tools, e.g., root loci or Bode plots, to study the impact of parameters.

8.3.1 Small-signal model for the mechanical system

Oscillatory torsional modes cannot be adequately modeled using a single mass to represent the rotor shaft. In this simple example, a two-mass rotor is modeled to demonstrate torsional interaction. The purpose of this example is to demonstrate the interaction of rotor torsional modes and the LC resonance modes. The parameters used are for demonstration only and are not aligned with any real system.

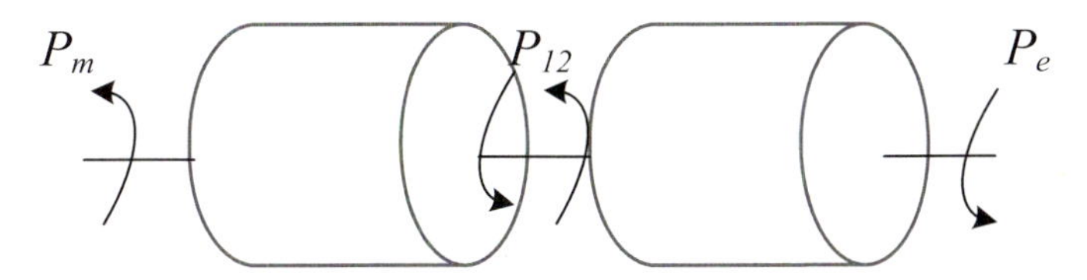

Figure 8.11: Two-mass rotor shaft.

If we assume that the mechanical system has two masses as shown in

Figure 8.11, then the overall block diagram for the two-mass system is shown in Figure 8.12.

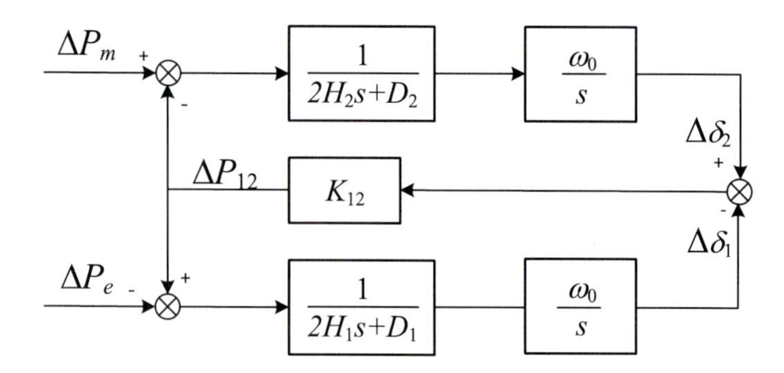

Figure 8.12: Two-mass rotor shaft block diagram.

$$\Delta\delta_1 = \frac{1}{2H_1 s + D_1} \frac{\omega_0}{s} (K_{12}(\Delta\delta_2 - \Delta\delta_1) - \Delta P_e) \tag{8.53a}$$

$$\Delta\delta_2 = \frac{1}{2H_2 s + D_2} \frac{\omega_0}{s} (\Delta P_m - K_{12}(\Delta\delta_2 - \Delta\delta_1)) \tag{8.53b}$$

where H_i and D_i are the inertia and damping coefficient related to ith mass in the rotor. P_{12} is the power related to the angle displacement of the two rotor masses, $\Delta P_{12} = K_{12}(\Delta\delta_2 - \Delta\delta_1))$, K_{12} is a coefficient.

Adding the two equations after multiplying $(2H_1 s + D_1)$ to both sides of (8.53a) and multiplying $(2H_2 s + D_2)$ to both sides of (8.53b) leads to

$$(2H_1 s + D_1)\Delta\delta_1 + (2H_2 s + D_2)\Delta\delta_2 = \frac{\omega_0}{s}(\Delta P_m - \Delta P_e).$$

Therefore the expression of $\Delta\delta_2$ can be expressed in terms of $\Delta\delta_1$, ΔP_m and ΔP_e.

$$\Delta\delta_2 = \frac{1}{2H_2 s + D_2} \left(\frac{\omega_0}{s}(\Delta P_m - \Delta P_e) - (2H_1 s + D_1) \right). \tag{8.54}$$

Substituting $\Delta\delta_2$ by (8.54) in (8.53a) leads to the transfer function matrix from $[\Delta P_m, \Delta P_e]^T$ to $\Delta\delta_2$.

$$\Delta\delta_1 = \begin{bmatrix} G_1 & -G_2 \end{bmatrix} \begin{bmatrix} \Delta P_m \\ \Delta P_e \end{bmatrix} \tag{8.55}$$

where $G_1 = \dfrac{\omega_0 K_{12}}{D(s)}$

$$G_2 = \frac{\omega_0 K_{12} + s(2H_2 s + D_2))}{D(s)}$$

$$D(s) = s\left(s\frac{(2H_1 s + D_1)(2H_2 s + D_2)}{\omega_0} + K_{12}(2H_1 s + D_1 + 2H_2 s + D_2)\right)$$

The poles and zeros of G_2 are shown in Figure 8.13. As a comparison, the transfer function for a one-mass mechanical system $\frac{\omega_0}{s(2H_1 s + D_1)}$ has two poles on the real-axis: 0, and $-\frac{D_1}{2H_1}$. We can see that compared to the one-mass system, the two-mass system introduced a pair of additional complex conjugate poles and a pair of complex conjugate zeros.

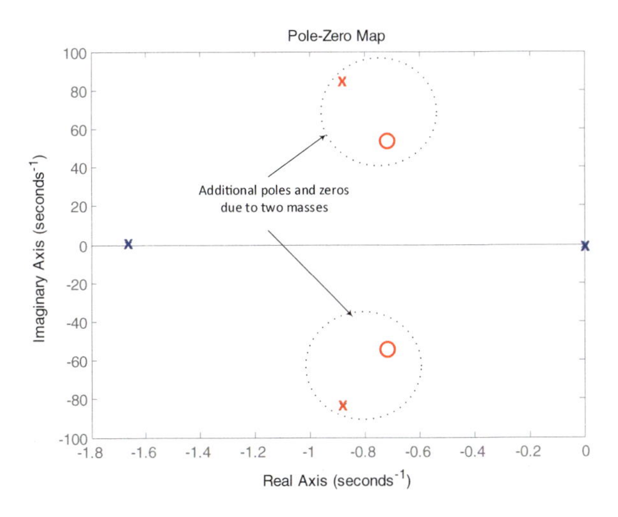

Figure 8.13: Poles and zeros of G_2.

8.3.2 Small-signal model for complex power for an RLC circuit

The above paragraph gives the transfer function related ΔP_e to $\Delta\delta_1$ in the mechanical system. The subscript 1 will be dropped for δ_1 in the following paragraphs. Next, we seek the relationship from $\Delta\delta$ to ΔP_e of the electric network. The transfer function is notated as J. The closed-loop system is shown as Figure 8.14.

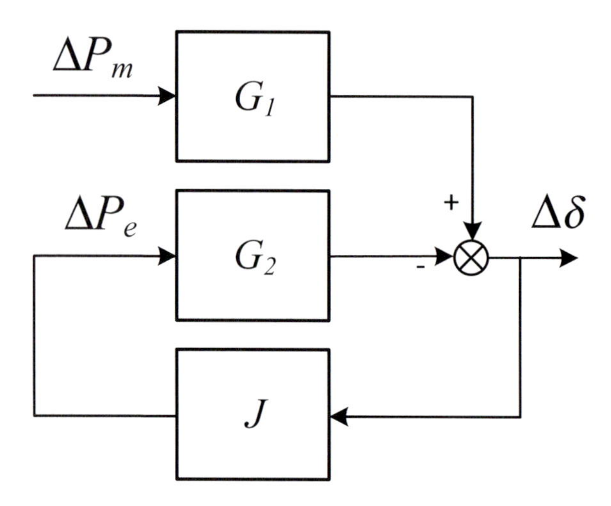

Figure 8.14: System block diagram.

If we know J, we can judge system stability using the loop gain JG_2. The closed-loop system poles are located at the root loci when the gain is 1. J will be found by examining the complex power expression for the RLC circuit in Figure 8.15.

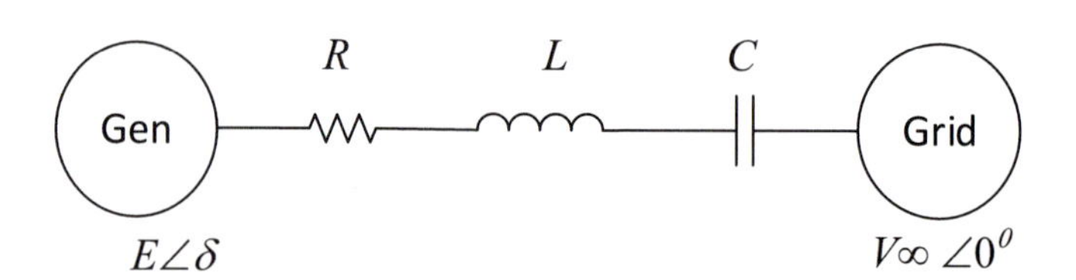

Figure 8.15: An RLC circuit.

In the synchronous rotating reference frame, the complex vectors of the generator internal voltage and the grid voltage are as follows.

$$\bar{E} = E\angle\delta, \quad \bar{V}_\infty = V_\infty\angle 0 \tag{8.56}$$

where V_∞ is assumed to be constant.

The electromagnetic dynamics of the line will not be ignored. Therefore, in Laplace domain, the current and voltage relationship in the dq reference frame is

$$\bar{I}(s) = \frac{\bar{E}(s) - \bar{V}_\infty(s)}{R + (s + j\omega_1)L + \frac{1}{(s+j\omega_1)C}} \quad (8.57)$$

$$\Delta\bar{I}(s) = \frac{\Delta\bar{E}(s)}{R + (s + j\omega_1)L + \frac{1}{(s+j\omega_1)C}} \quad (8.58)$$

Note that the line impedance model in *abc* reference frame is $R + sL + \frac{1}{sC}$, while in a synchronous reference frame it becomes $R+(s+j\omega_1)L+\frac{1}{(s+j\omega_1)C}$, where ω_1 is the fundamental component's frequency.

We replace s of a transfer function based on the static reference frame by $s + j\omega_1$. The resulting transfer function is based on the synchronous reference frame rotating at ω_1. This relationship is explained as follows.

A space vector and a complex vector based on the synchronous reference frame have the following relationship:

$$\vec{f}(t) = \overline{F}(t)e^{j(\omega_1 t+\theta_0)}, \quad (8.59)$$

where θ_0 is the initial angle between the rotating reference frame and the static reference frame when $t = 0$.

In a frequency domain, then the relationship between their corresponding Laplace transforms is:

$$\vec{f}(s) = \overline{F}(s - j\omega_1). \quad (8.60)$$

This is the same as:

$$\vec{f}(s + j\omega_1) = \overline{F}(s). \quad (8.61)$$

Therefore, if we know a transfer function's expression in the static frame, to find its expression in a rotating reference frame, we just need to replace s by $s + j\omega$ where ω is the rotating reference frame's speed.

Since the complex power from the generator can be expressed as $S = \bar{E}\bar{I}^*$, its small-signal expression will be:

$$\Delta S = \bar{E}\Delta\bar{I}^* + \bar{I}^*\Delta\bar{E} \quad (8.62)$$

The above derivation is not trivial. Small-signal expression has the following basic assumption. At steady-state, the values of the variables should be constant. If instantaneous voltages and currents are used, e.g., v_{abc} and i_{abc}, we can not derive small-signal expressions for Δv_{abc} or Δi_{abc} since at steady-state the voltages and currents are periodic.

Therefore, it is critical to express the complex power by voltages and currents in the dq reference frame. At steady-state, voltages and currents in dq reference frames are constants.

Substituting $\Delta\bar{I}(s)$ in the above equation by (8.58), also considering that $\Delta\bar{E} = Ee^{j\delta}j\Delta\delta = j\bar{E}\Delta\delta$, we have

$$\Delta\bar{E}^* = -j\bar{E}^*\Delta\delta \tag{8.63}$$

$$\Delta S = \bar{E}\frac{\Delta\bar{E}^*}{Z^*} + \Delta\bar{E}\bar{I}^* \tag{8.64}$$

$$= \left(-\frac{jE^2}{Z^*} + j\bar{E}\bar{I}^*\right)\Delta\delta \tag{8.65}$$

$$= \left(-\frac{jE^2}{Z^*} + j(P_e + jQ_e)\right)\Delta\delta \tag{8.66}$$

where $Z = R + (s + j\omega_1)L + \frac{1}{(s+j\omega_1)C}$.

Let $G = \frac{1}{Z^*} = G_R + jG_I$, we then have

$$\Delta S = [-jE^2(G_R + jG_I) + j(P_e + jQ_e)]\Delta\delta \tag{8.67}$$

$$\Delta P_e = (G_I E^2 - Q_e)\Delta\delta \tag{8.68}$$

We find the transfer function from $\Delta\delta$ to ΔP_e due to the electric network J as the following.

$$J = G_I E^2 - Q_e \tag{8.69}$$

where

$$G_I = \frac{\omega_1 C(s^2 + \omega_1^2)(L^2(s^2 + \omega_1^2) - 1)}{(C(R + sL)(s^2 + \omega_1^2) + s)^2 + \omega_1^2(1 - LC(s^2 + \omega_1^2))^2} \tag{8.70}$$

If $Q_e = 0$ and $E = 1$, then $J = G_I$. J has two pairs of complex conjugate poles and one pair of complex conjugate zeros.

8.3.3 Stability analysis

It is obvious that the poles and zeros are dependent on the value of C or the series compensation degree. Figure 8.16 shows the poles and zeros related to J as well as the poles and zeros related to G_2. With the compensation degree increasing, the poles and zeros of J move closer to the original point.

Figure 8.16 shows the root loci of the loop gain JG_2. It can be seen that when the compensation degree is 80%, the closed-loop system loses stability since when the gain is 0.526, two root loci move to the RHP. This

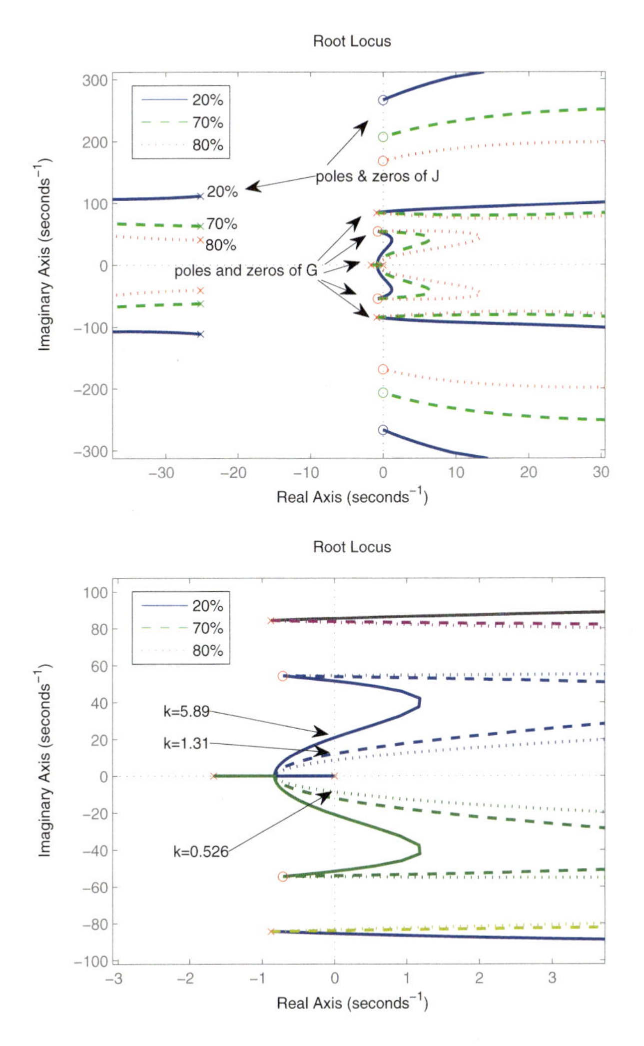

Figure 8.16: Torsional interaction due to a high series compensation. The root loci of JG_2 are plotted. It can be seen that when the compensation degree is 80%, the closed-loop system loses stability since when the gain is 0.526, two root loci move to the right-half-plane.

indicates instability at 80% compensation degree. If the mechanical system is modeled by a one-mass system, instability will not happen. Thus, this instability is termed the torsional interaction of the mechanical system with the RLC circuit.

Step responses of $1/(1+JG_2)$ at three compensation levels are shown in

Figure 8.17. It can be seen that at 20% and 70%, the system is stable, while at the 80% compensation level, the system loses stability. The oscillation frequency is about 2 Hz. The close-loop system roots can be found when $k = 1$ from the root loci plot in Fig 8.16. The corresponding roots are $0.812 \pm j12$. The imaginary part corresponds to an oscillation frequency of $\frac{12}{2\pi} = 1.91$ Hz. The analysis based on root loci corroborates the linear system simulation results.

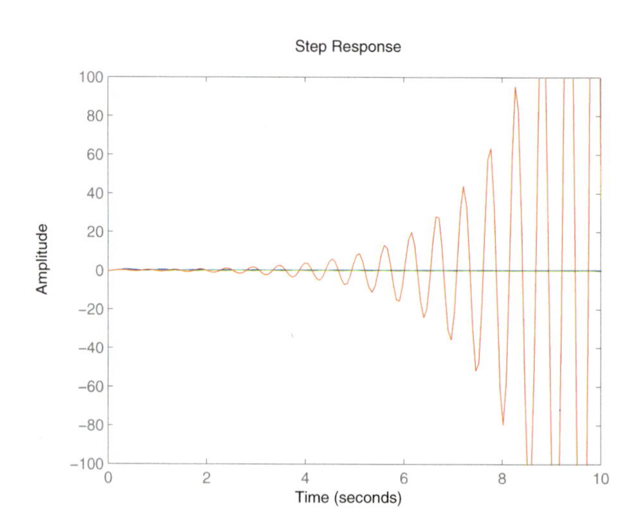

Figure 8.17: Torsional interaction at 80% compensation degree.

References

Anderson, P. M. and A. A. Fouad (2008). *Power System Control and Stability.* John Wiley & Sons.

Aström, K. J. and R. M. Murray (2010). *Feedback Systems: An Introduction for Scientists and Engineers.* Princeton University Press.

Bergen, A. R. and V. Vittal (2009). *Power Systems Analysis.* Pearson Education India.

Chandorkar, M. C., D. M. Divan, and R. Adapa (1993). Control of parallel connected inverters in standalone AC supply systems. *IEEE Transactions on Industry Applications 29*(1), 136–143.

Chapra, S. C. and R. P. Canale (2012). *Numerical Methods for Engineers,* Volume 2. McGraw-Hill New York.

Chow, J. H. and K. W. Cheung (1992). A toolbox for power system dynamics and control engineering education and research. *IEEE transactions on Power Systems 7*(4), 1559–1564.

Crow, M. L. (2015). *Computational Methods for Electric Power Systems.* CRC Press.

Dorf, R. C. and R. H. Bishop (1998). *Modern Control Systems.* Pearson (Addison-Wesley).

Fan, L. (2017). Interarea oscillations revisited. *IEEE Transactions on Power Systems 32*(2), 1585–1586.

Fax, J. A. and R. M. Murray (2004). Information flow and cooperative control of vehicle formations. *IEEE Transactions on Automatic Control 49*(9), 1465–1476.

Fortescue, C. L. (1918). Method of symmetrical co-ordinates applied to the solution of polyphase networks. *Transactions of the American Institute of Electrical Engineers 37*(2), 1027–1140.

Gabor, D. (1946). Theory of communication. part 1: The analysis of information. *Journal of the Institution of Electrical Engineers-Part III: Radio and Communication Engineering 93*(26), 429–441.

Hengster– Movric, K., F. L. Lewis, and M. Sebek (2015). Distributed static output-feedback control for state synchronization in networks of identical LTI systems. *Automatica 53*, 282–290.

Heydt, G., S. Venkata, and N. Balijepalli (2000). High impact papers in power engineering, 1900-1999. In *North American Power Symposium, University of Waterloo, Waterloo, Canada, pp 1-7.*

Hingorani, N. G. and L. Gyugyi (2000). *Understanding FACTS: Concepts and Technology of Flexible AC Transmission Systems.* Wiley-IEEE Press.

Kosterev, D. N., C. W. Taylor, and W. A. Mittelstadt (1999). Model validation for the august 10, 1996 wscc system outage. *IEEE Transactions on Power Systems 14*(3), 967–979.

Krause, P. (1986). *Analysis of Electric Machinery.* New York: McGraw-Hill.

Kundur, P., N. J. Balu, and M. G. Lauby (1994). *Power System Stability and Control,* Volume 7. McGraw-Hill New York.

Li, Y. W. and C.-N. Kao (2009). An accurate power control strategy for power-electronics-interfaced distributed generation units operating in a low-voltage multibus microgrid. *IEEE Transactions on Power Electronics 24*(12), 2977–2988.

Miao, Z. and L. Fan (2017, jan). Achieving economic operation and secondary frequency regulation simultaneously through feedback control. *IEEE Transactions on Power Systems 32*(1), 85–93.

Mohan, N. and T. M. Undeland (2007). *Power Electronics: Converters, Applications, and Design.* John Wiley & Sons.

North American Electric Reliability Corporation (2011). Interconnection criteria for frequency response requirements.

Park, R. H. (1929). Two-reaction theory of synchronous machines generalized method of analysis-Part I. *Transactions of the American Institute of Electrical Engineers 48*(3), 716–727.

Rogers, G. (2012). *Power System Oscillations.* Springer Science & Business Media.

Walker, D. N., C. E. J. Bowler, R. L. Jackson, and D. A. Hodges (1975, Sept). Results of subsynchronous resonance test at Mohave. *IEEE Transactions on Power Apparatus and Systems 94*(5), 1878–1889.

Yazdani, A. and R. Iravani (2010). *Voltage-Sourced Converters in Power Systems: Modeling, Control, and Applications.* John Wiley & Sons.

Index